Springer Theses

Recognizing Outstanding Ph.D. Research

For further volumes:
http://www.springer.com/series/8790

Aims and Scope

The series "Springer Theses" brings together a selection of the very best Ph.D. theses from around the world and across the physical sciences. Nominated and endorsed by two recognized specialists, each published volume has been selected for its scientific excellence and the high impact of its contents for the pertinent field of research. For greater accessibility to non-specialists, the published versions include an extended introduction, as well as a foreword by the student's supervisor explaining the special relevance of the work for the field. As a whole, the series will provide a valuable resource both for newcomers to the research fields described, and for other scientists seeking detailed background information on special questions. Finally, it provides an accredited documentation of the valuable contributions made by today's younger generation of scientists.

Theses are accepted into the series by invited nomination only and must fulfill all of the following criteria

- They must be written in good English.
- The topic should fall within the confines of Chemistry, Physics, Earth Sciences, Engineering and related interdisciplinary fields such as Materials, Nanoscience, Chemical Engineering, Complex Systems and Biophysics.
- The work reported in the thesis must represent a significant scientific advance.
- If the thesis includes previously published material, permission to reproduce this must be gained from the respective copyright holder.
- They must have been examined and passed during the 12 months prior to nomination.
- Each thesis should include a foreword by the supervisor outlining the significance of its content.
- The theses should have a clearly defined structure including an introduction accessible to scientists not expert in that particular field.

Rituparna Bose

Biodiversity and Evolutionary Ecology of Extinct Organisms

Doctoral Thesis accepted by
Indiana University, Bloomington,
United States of America

 Springer

Author
Dr. Rituparna Bose
City College of New York
New York, NY
USA

Supervisor
Prof. Dr. P. David Polly
Department of Geological Sciences Office
Indiana University
Bloomington, IN
USA

ISSN 2190-5053   ISSN 2190-5061   (electronic)
ISBN 978-3-642-44059-5   ISBN 978-3-642-31721-7 (eBook)
DOI 10.1007/978-3-642-31721-7
Springer Heidelberg New York Dordrecht London

Parts of this thesis have been published in the following journal articles:

Bose, R., Schneider, C., Leighton, L. R., and Polly, P. D., 2011. Influence of atrypid morphological shape on Devonian episkeletobiont assemblages from the lower Genshaw Formation of the Traverse Group of Michigan: a geometric morphometric approach. *Paleogeography, Paleoecology and Paleoclimatology* 310:427–441.

Bose, R., 2012. A new morphometric model in distinguishing two closely related extinct brachiopod species. Manuscript accepted in *Historical Biology: An International Journal of Paleobiology* 24:1–10 (DOI:10.1080/08912963.2012.658568)

I would like to dedicate this work to Professor David Polly, a true pioneer in the field of Evolutionary Paleontology and geometric morphometrics and a great source of inspiration for me in this field of science; I would also like to dedicate this work to my family who were a great source of moral support throughout my doctoral career

Supervisor's Foreword

Organisms are trade-offs: their structure neither perfectly suits their needs, nor is it random. The vertebrae in our own lower backs are structured differently from our four-legged relatives, providing us with support against gravity for our upper bodies and flexibility for forward locomotion. Despite the transformation in vertebral structure that accompanied the evolution of bipedality, our lower backs are notoriously rickety, subjecting approximately three quarters of us to severe back pain sometime during our adult lives. The structure of our lower back simply is not adequate to both hold our skulls high and move us forward smoothly on our two legs. Like all organismal structures, our backs are a compromise between what we inherited from our ancestors, adaptation of that inheritance through the painful process of natural selection, and non-genetic compensation for stress, strain, and breakage provided by growth and remodeling. Each one of these factors packs its own set of compromises. Our bones are both rigid structural supports whose hydroxyapatite matrix requires a calcium-rich diet and a mineral reservoir to provide our nervous systems with calcium when dietary sources are low. These two functions that are so important to our health often work at cross purposes as anyone with osteoporosis can attest. Every step in the Darwinian descent with modification is a response to the many competing selectional demands of the moment and chance.

The study of the factors that influence organismal structure is both fascinating and challenging. The trade-offs mean that no one factor is ever expected to explain all of the variations we see in a structure; indeed even important factors may only explain a small portion of that variation. The study of the evolution of structures is therefore a statistical pursuit. In this dissertation, Dr Rituparna Bose uses a statistical approach to look at the evolution of brachiopod valves. Brachiopods were important members of marine communities during the Paleozoic Era (542–251 million years ago), living through many cycles of sea level change and extensive reorganization of the continental shelf environments in which they lived. Bose uses a series of case studies of the Atrypidae, a widespread group of brachiopods, to examine the roles of ancestry (phylogeny), geographic isolation, sea level change, encrustations by other marine organisms, and adaptations to substrates as factors explaining the shape of the valves of these filter-feeding animals. She uses geometric morphometrics, a relatively recent method for measuring biological shape, to analyze the shape of features of the valves and to statistically determine the extent to which these factors are each correlated with valve shape. She finds that despite considerable similarities among atrypides through time and space, their shell shape is indeed a measureable product of ancestry and environmental change.

Prof. Dr. P. David Polly
Associate Professor of Geological Sciences
Paleontology, Indiana University, Bloomington

About the Author

Dr. Rituparna Bose (M.Sc., Ph.D.) is currently an adjunct Assistant Professor in the Department of Earth and Atmospheric Sciences and in the Department of Biological Sciences and Geology at the City University of New York.

Dr. Bose obtained her undergraduate education at the University of Calcutta, India, and authored a thesis paper in environmental geology (coastal management). She came to the United States in 2004 to pursue higher studies. During the course of her graduate studies at Bowling Green State University and Indiana University at Bloomington, she was the primary author in multiple high-impact peer-reviewed publications. Here she was awarded the Indiana University Dissertation Year Research Fellowship which is given to the best doctoral students of the university. Her work with Dr. Margaret Yacobucci and Prof. David Polly culminated in major findings in evaluating the evolution and determining the biodiversity of extinct organisms.

As a result of her findings which have profound implications in bio-conservation, she won major national awards like the Theodore Roosevelt Memorial Grant (American Museum of Natural History) and Schuchert and Dunbar Grant (Yale Peabody Museum of Natural History). Additionally, BP Global Energy Group funded her to present these findings at North American Paleontology Convention (NAPC) by the prestigious NAPC Travel Award.

Dr. Bose continues to pursue both her teaching and research career. Her research interests lie in applications of quantitative algorithms to study evolutionary biology, micropaleontology, disaster management studies, and other applied geological sciences. In addition, she teaches courses in Physical Geology, Environmental Geology, Earth System Science, Natural Disasters, and Historical Geology.

The foreword of this book has been written by Prof. David Polly, Editor of *Palaeontology* and Executive Editor of *Palaeontologia Electronica*.

Acknowledgments

First and foremost, I would like to express my deep gratitude to my supervisor, Prof. David Polly, for his valuable guidance. I would also like to gratefully acknowledge Prof. Claudia Johnson for her valuable suggestions in various professional aspects throughout my graduate education. Finally, I would like to specially thank Prof. Lindsey Leighton and Prof. Chris Schneider at the University of Alberta for their kind help.

Financial support for this research was derived primarily from Galloway-Horowitz Research Grant-in-Aid granted by Department of Geological Sciences, Indiana University; Indiana University School of Arts and Sciences Dissertation Year Research Fellowship; Dunbar and Schuchert Grant-in Aid funded by Yale Peabody Museum, Theodore Roosevelt Memorial Grant funded by American Museum of Natural History, and funds from the BP Global Energy group.

Contents

Chapter 1
A Geometric Morphometric Approach in Assessing Paleontological Problems in Atrypid Taxonomy, Phylogeny, Evolution and Ecology

1.1 Introduction

1.1.1 Brief Outline

This section is a brief overview of the dissertation. A brief outline of all the dissertation chapters and a few major proposed questions relevant to each chapter is discussed here.

Chapter 2 follows this Introduction with a detailed morphometric shape analysis of taxonomic and phylogenetic arrangement of atrypid subfamilies from a prior proposed phylogeny by Copper (1973), (Bose et al. 2011a, b). Do morphological shape distances between genera comply with the phylogenetic chart proposed by Paul Copper? Do pairwise Procrustes distances between generic pairs agree with the taxonomic arrangement in atrypid subfamilies as proposed by Copper (1973) in the past? Overall, do these genera reflect stasis within the P3 EEU? Chapter 3 presents tests for stasis within an atrypid species lineage, *Pseudoatrypa cf. lineata* from the middle Devonian Traverse Group of Michigan, over a period of 5 m.y. period (Bose and Polly 2011). Does the morphology of this species remain static within the Hamilton EESU stratigraphic units? What could be the possible causes (environment or evolution?) behind the morphological pattern observed? Do we see similar morphological patterns in species lineages in the Traverse Group of Michigan Basin when compared with those of the Hamilton Group of Appalachian Basin? Chapter 4 presents a detailed analysis of episkeletobiont interactions with the Genshaw Formation atrypids from the middle Devonian Traverse Group, with further investigation of species, valve and location preference of these episkeletobiont assemblages. Can we reconstruct life orientation of brachiopods from location preferences of certain episkeletobiont associations on their host valves? Can morphological variation in host species influence the rate of encrustation?

R. Bose, *Biodiversity and Evolutionary Ecology of Extinct Organisms*,
Springer Theses, DOI: 10.1007/978-3-642-31721-7_1,

In addition to providing information about factors of evolution, the response of valve morphology to both its physical environment and community interactions is important in reconstructing life histories. Studying the morphology of extinct brachiopod groups, thus, will help reconstruct their evolutionary history in terms of both large scale and small scale temporal intervals. The study presented in this dissertation is important because (a) the Silurian-Devonian is an EEU that has frequently been scrutinized for stasis, (b) the middle Devonian Traverse Group is an EESU which has not been tested for stasis, (c) atrypides are among the most diverse and common macro-invertebrates in these intervals, but (d) with the exception of one athyride and one spiriferide species from the middle Devonian Hamilton Group in the Lieberman et al. (1995) study, the use of morphometrics to evaluate stasis in atrypides is virtually uninvestigated, (e) morphological variation in Devonian brachiopod species *Pseudoatrypa cf. lineata* and their influence on episkeletobiont assemblages using morphometrics has not been examined in the past, and (f) finally, lophophores are the primary feeding and respiratory organ in these organisms and determining their morphological shape could help predict the lophophore shape of these extinct organisms.

1.1.2 Why Silurian and Devonian Time Periods for this Study?

After the Ordovician mass extinction event, many marine taxa rediversified in the Late Ordovician event and a few persisted in the Silurian and Devonian ecosystem. Trilobites almost disappeared, bivalve mollusks invaded non-marine habitats, corals and stromatoporoids diversified in new ways giving rise to massive reef systems in shallow seas, and graptolites diversified to a great extent. Brachiopods attained the highest diversity during this time. New forms like ammonoids and jawed fishes flourished for the first time. Vascular plants invaded the land in late Silurian period followed by the evolution of more complex land plants in the early and middle Devonian period. Arthropods (insects, scorpions and spiders) and vertebrate animals also evolved during this period. During the Devonian, spore plants were accompanied by seed plants and large trees with roots and abundant foliage arose in the Late Devonian. This initial spread of terrestrial vegetation accelerated weathering rates thus resulting in relatively cooler climates during this time, further providing shelter for early vertebrates (Ausich and Lane 1999; Stanley 2005).

The Late Ordovician period was marked by a significant change in the sedimentation pattern in Eastern North America when the newly formed mountains in the east from the Taconic orogenic event caused the deposition of clastic wedges in the west, as reviewed by Stanley (2005). The pattern continued in the Silurian period followed by erosion of the eastern mountains and inundation of the clastic deposits by the shallow epicontinental seas. During the Late Silurian time, shallow water carbonate sediments accumulated along a new passive margin. Carbonate sedimentation that spread along the continent during the Devonian had an abrupt subsidence and was soon replaced by a foreland basin as mountains rose in the east (Acadian orogeny initiation during middle Silurian). During most of the Devonian, eastern North America accumulated little or no sedimentation. The carbonate platforms were covered with sandy beach deposits before subsidence, and then later after subsidence, black muds, turbidites, siltstones and shales deposited in Hudson foreland (New York) basin near the end Devonian time. Shallow water sedimentation then lead to the deposition of deep water flysch. Due to an enhanced supply of sedimentation from erosion of adjacent mountain belts in the late Devonian time, these deep water deposits in the foreland basin finally gave way to the deposition of shallow marine and non-marine molasse deposits.

To the west of North America, patterns of sedimentation, however, changed during Early Silurian. The Michigan and Ohio basins accumulated muddy carbonates and were well populated with patchy reefs and bounded by large barrier reefs. The east continued to be filled in with siliciclastic muds. The overall pattern changed with the progression of time; during the Silurian time, siliciclastic sedimentation slowly gave way to carbonate sedimentation in the east. The barrier reefs around the northeastern (Michigan and Ohio) basins had risen very high restricting the supply of water to the basins. This, together with lowering of sea level, caused precipitation of evaporitic deposits in the margins and at considerable depths of the basins. During the Late Silurian, reefs grew only in the southwest (Indiana and Illinois), given the unfavorable conditions in the evaporitic basins. Later in the Devonian time, black mud deposition extended to the west, covering a vast area of eastern and central North America with these sediments, further eliminating nearly all members of the coral-stromatoporid reef community and the placoderm fishes (Stanley 2005).

1.1.3 Why Brachiopods?

Brachiopods are excellent models for testing macroevolutionary and ecological hypotheses due to their increased abundance and diversity during the middle Paleozoic interval after the great Ordovician biodiversification event (Alexander 1986; Jin 2001; Leighton 2005; Novack-Gottshall and Lanier 2008). Their slow growth rate, low metabolism, obligate filter feeding, restriction to hard substrates and low energy environments, suggest greater chances of their survival from climatic changes, more likely than any other invertebrate taxa (Rudwick 1970; Thayer 1977; Peck 2008). Brachiopod morphology has been studied using advanced morphometric and statistical techniques in the past two decades (Goldman and Mitchell 1990; Lieberman et al. 1995; Krause 2004; Tomasovych et al. 2008; Bose et al. 2010, 2011a, b), further depicting the shape differences at the community and species level.

For this study, atrypides have been selected as with their origination in the late Middle Ordovician (Llandeilo) time, they persisted for sometime in the Silurian and Devonian time periods with high generic diversity and abundance, until they were finally wiped out during the late Devonian mass extinction (Frasnian) (Copper 2001a, b). Atrypides were extensively studied in the past by prior researchers (Fenton and Fenton 1932, 1935; Biernat 1964; Copper 1967, 1973, 1977, 1982, 1995, 1997, 2001a, b, 2002, 2004; Day 1995, 1998; Day and Copper 1998; Ma and Day 2007), but to date no one has applied advanced quantitative techniques to further address evolutionary, ecological, taphonomic, taxonomic and phylogenetic inquiries. This study was a first attempt in resolving such aspects of paleontological problems by using geometric morphometric methods.

Episkeletobionts, also known as encrusting organisms that adhere to the surface of a shell (Taylor and Wilson 2002), were mostly restricted to hard skeletons of the host shells during the Devonian geologic period (Taylor and Wilson 2003). Episkeletobionts on brachiopod hosts have been extensively studied in the past (Rudwick 1962; Richards and Shabica 1969; Richards 1972; Hurst 1974; Thayer 1974; Pitrat and Rogers 1978; Kesling et al. 1980; Spjeldnaes 1984; Bordeaux and Brett 1990; Rodland et al. 2004, 2006; Schneider and Leighton 2007; Rodrigues et al. 2008; Bose et al. 2010; 2011a, b). Thus, quantitative interpretation of abundance, diversity and location preference of these Devonian episkeletobionts on host brachiopod valves could provide insight into the paleoecology and life orientation of these brachiopod hosts.

1.1.4 Why Geometric Morphometric Methods?

Geometric morphometrics provides a great insight into the biological, paleoecological and evolutionary processes that affects the morphology of organisms. Geometric morphometric methods are helpful in solving many complex hypotheses in shape comparative studies. In the past two decades, new methods biological and geometric shape analysis have been elucidated (Bookstein 1991; Zelditch et al. 2004), and applied in solving evolutionary problems in numerous fields. These techniques have been applied to problems in ontogeny and phylogeny (Fink and Zelditch 1995; Adams and Rosenberg 1998; Rohlf 1998; MacLeod 1999, 2001, 2002; Lockwood et al. 2004; Cardini and O'Higgins 2004; Rook and O'Higgins 2005; Caumul and Polly 2005), hybridization (Hayden et al. 2010), functional morphology (Bonnan 2004; Kassam et al. 2004; Stayton 2006; Kulemeyer et al. 2009), genetics (Myers et al. 2006), primate and human cranial anatomy (Frost et al. 2003; Couette et al. 2005; Bernal et al. 2006), mammalian evolution (Polly 2003, 2004, 2005, 2007), trilobite evolution, paleoecology and taphonomy (Webster and Hughes 1999; Webber and Hunda 2007; Webster and Zelditch 2011) and brachiopod ontogeny, evolution, systematics and paleoecology (Haney et al.

2001; Krause 2004; Tomasovych et al. 2008; Bose et al. 2010, 2011a, b). These references provide a great overview of applications, while those interested in more detailed technical explanations are referred to the primary reference (Bookstein 1991; Adams et al. 2004). Collectively, this new set of methods for analyzing landmark-based data is referred to as geometric morphometrics.

Why use geometric morphometrics over any other technique for assessing morphological change in time and space? Geometric morphometrics is an advanced analytical technique that has several advantages over traditional morphometric techniques involving the measurement of interlandmark distances, angles or ratios of distances (Rohlf 1990). The direct use of landmark data in the analysis allows the incorporation of a unique feature of those data, their spatial organization and the inclusion of average forms, adding statistical power to the various shape space components, allowing pure shape to be analyzed independent of size and allowing results to be depicted visually with representations that resemble the original objects (Bookstein 1991).

References

Adams DC, Rosenberg MS (1998) Partial-warps, phylogeny, and ontogeny: a comment on Fink and Zelditch (1995). Sys Biol 47:168–173

Adams DC, Rohlf FJ, Slice DE (2004) Geometric morphometrics: ten years of progress following the 'revolution'. Italian J Zool 71:5–16

Alexander RR (1986) Resistance to and repair of shell breakage induced by durophages in late Ordovician brachiopods. J Paleontol 60:273–285

Ausich WI, Lane NG (1999) Life of the past. Prentice Hall, New Jersey

Bernal V, Perez SI, Gonzalez PN (2006) Variation and causal factors of craniofacial robusticity in Patagonian hunter-gatherers from the late Holocene. Am J Hum Biol 18:748–765

Biernat G (1964) Middle devonian atrypacea (brachiopoda) from the Holy Cross Mountains, Poland. Acta Palaeontol Pol 9:277–356

Bookstein FL (1991) Morphometric tools for landmark data: geometry and biology. Cambridge University Press, Cambridge

Bonnan MF (2004) Morphometric analysis of humerus and femur shape in Morrison sauropods: implications for functional morphology and paleobiology. Paleobiology 30(3):444–470

Bordeaux YL, Brett CE (1990) Substrate specific associations of epibionts on middle devonian brachiopods. Implic Paleoecol Hist Biol 4:203–220

Bose R, Schneider C, Polly PD, Yacobucci MM (2010) Ecological interactions between *Rhipidomella* (orthides, brachiopoda) and its endoskeletobionts and predators from the Middle Devonian Dundee Formation of Ohio. U. S. Palaios 25:196–210

Bose R, Schneider C, Leighton LR, Polly PD (2011a) Influence of atrypid morphological shape on Devonian episkeletobiont assemblages from the lower GENSHAW formation of the Traverse Group of Michigan. Geom Morphometr Approach Palaeogeogr, Palaeoclimatol, Palaeoecol 310:427–441

Bose R, Leighton LR, Day J, Polly PD (2011) A geometric morphometric approach in testing the taxonomy and phylogeny of eastern North American atrypid brachiopods. Manuscript in preparation for submission to Lethaia

Bose R, Polly PD (2011) Morphological evolution in *Pseudoatrypa cf. lineata* from the middle Devonian traverse Group of Michigan. Manuscript in preparation for submission to Paleobiology

Caumul R, Polly PD (2005) Phylogenetic and environmental components of morphological variation: Skull, mandible, and molar shape in marmots (Marmota, Rodentia). Evolution 59:2460–2472

Cardini A, O'Higgins P (2004) Patterns of morphological evolution in Marmota (Rodentia, Sciuridae): geometric morphometrics of the cranium in the context of marmot phylogeny, ecology and conservation. Biol J Linn Soc 82:385–407

Copper P (1967) Adaptations and life habits of Devonian atrypid brachiopods. Palaeogeogr Palaeoclimatol Palaeoecol 3:363–379

Copper P (1973) New Siluro-Devonian atrypoid brachiopods. J Paleontol 47:484–500

Copper P (1977) The Late Silurian brachiopod genus atrypoidea. Geologiska Foreningers Stockholm Forhandligas 99:10–26

Copper P (1982) Early silurian atrypoids from Manitoulin Island and Bruce Peninsula, Ontario. J Paleontol 56:680–702

Copper P (1990) Evolution of the atrypid brachiopods. In: MacKinnon DI, Lee DE, Campbell JD (eds) Brachiopod through time: proceedings of the second international brachiopod congress, Dunedin, New Zealand, pp 33–39

Copper P (1995) Five new genera of late Ordovician-early silurian brachiopods from Anticosti Island. East Can J Paleontol 69:846–862

Copper P (1997) New and revised genera of Wenlock-Ludlow Atrypids (silurian brachiopoda) from Gotland, Sweden, and the United Kingdom. J Paleontol 70:913–923

Copper P (2001a) Radiations and extinctions of atrypide brachiopods: Ordovician–Devonian. In: Brunton CHC, Cocks LRM, Long SL (eds) Brachiopods past and present. Natural History Museum, London, England, pp 201–211

Copper P (2001b) Reefs during the multiple crises towards the Ordovician-Silurian boundary: Anticosti Island, eastern Canada, and worldwide. Can J Earth Sci 38:153–171

Copper P (2002) Atrypida. In: Kaesler RL (ed) Brachiopoda (revised), part H of treatise on invertebrate paleontology. The Geological Society of America, Inc. and the University of Kansas, Boulder, Colorado and Lawrence, Kansas, pp. 1377–1474

Copper P (2004) Silurian (Late Llandovery-Ludlow) Atrypid brachiopods from Gotland, Sweden, and the Welsh Borderlands, the Great Britain. National Research Council of Canada Research Press, Ottawa

Couette S, Escarguel G, Montuire S (2005) Constructing, bootstrapping, and comparing morphometric and phylogenetic trees: a case study of new world monkeys (platyrrhini, primates). J Mammal 86:773–781

Day J (1995) Brachiopod fauna of the Upper Devonian (late Frasnian) Lime Creek Formation of north-central Iowa, and related units in eastern Iowa. In: Bunker BJ (ed) Geological society of Iowa guidebook, vol 62. pp. 21–40

Day J (1998) Distribution of latest givetian-frasnian atrypida (brachiopoda) in central and western North America. Acta Palaeontol Pol 43:205–240

Day J, Copper P (1998) Revision of latest givetian-frasnian atrypida (brachiopoda) from central North America. Acta Palaeontol Pol 43:155–204

Fenton CL, Fenton MA (1932) Orientation and injury in the genus Atrypa. Am Midl Nat 13:63–74

Fenton CL, Fenton MA (1935) Atrypae described by Clement L. Webster and related forms (Devonian, Iowa). J Paleontol 9:369–384

Fink WL, Zelditch ML (1995) Phylogenetic analysis of ontogenetic shape transformations: a reassessment of the piranha genus pygocentrus (teleostei). Sys Biol 44:344–361

Frost SR, Marcus LF, Bookstein FL (2003) Cranial allometry, phylogeography, and systematics of large-bodied papionins (Primates: Cercopithecinae) inferred from geometric morphometric analysis of landmark data. Anatomical record Part A-discoveries. Mol Cell Evol Biol 275:1048–1072

Goldman D, Mitchell CE (1990) Morphology, systematics, and evolution of middle devonian ambocoeliidae (brachiopoda), Western New York. J Paleontol 64:79–99

Haney RA, Mitchell CE, Kim K (2001) Geometric morphometric analysis of patterns of shape change in the Ordovician brachiopod *Sowerbyella*. Palaios 16:115–125

Hayden B, Pulcini D, Kelly-Quinn M, O'Grady M, Caffrey J, McGrath A, Mariani S (2010) Hybridisation between two cyprinid fishes in a novel habitat: genetics, morphology and life-history traits. Evol Biol 10

Hurst JM (1974) Selective epizoan encrustation of some silurian brachiopods from Gotland. Palaeontology 17:423–429

Jin J (2001) Evolution and extinction of the North American Hiscobeccus brachiopod fauna during the late Ordovician. Can J Earth Sci 38:143–151

Kassam D, Adams DC, Yamaoka K (2004) Functional significance of variation in trophic morphology within feeding microbabitat-differentiated cichlid species in Lake Malawi. Animal Biol 54:77–90

Kesling RV, Hoare RD, Sparks DK (1980) Epizoans of the Middle Devonian brachiopod *Paraspirifer bownockeri*: their relationships to one another and to their host. J Paleontol 54:1141–1154

Krause RA Jr (2004) An assessment of morphological fidelity in the sub-fossil record of a terebratulide brachiopod. Palaios 19:460–476

Kulemeyer C, Asbahr K, Gunz P, Frahnert S, Bairlein F (2009) Functional morphology and integration of corvid skulls—a 3D geometric morphometric approach. Front Zool 6:1–14

Leighton LR (2005) Comment—a new angle on strophomenid paleoecology: trace-fossil evidence of an escape Response for the plectambonitoid brachiopod *Sowerbyella rugosa* from a tempestite in the upper Ordovician kope formation (Edenian) of Northern Kentucky (Dattilo, 2004). Palaios 20:596–600

Lieberman BS, Brett CE, Eldredge N (1995) A study of stasis in two species lineages from the Middle Devonian of New York State. Paleobiology 21:15–27

Lockwood CA, Kimbel WH, Lynch JM (2004) Morphometrics and hominoid phylogeny: support for a chimpanzee-human clade and differentiation among great ape subspecies. Proc Nat Acad Sci U S A 101:4356–4360

Ma X, Day J (2007) Morphology and revision of late devonian (early famennian) *Cyrtospirifer* (brachiopoda) and related genera from South China and North America. J Paleontol 81:286–311

MacLeod N (1999) Generalizing and extending the eigen shape method of shape space visualization and analysis. Paleobiology 25:107–138

MacLeod N (2001) Landmarks, localization, and the use of morphometrics in phylogenetic analysis. In: Edgecombe G, Adrain J, Lieberman B (eds) Fossils, phylogeny, and form: an analytical approach. Kluwer Academic/Plenum, New York, pp 197–233

MacLeod N (2002) Testing evolutionary hypotheses with adaptive landscapes: use of random morphological simulation studies: Math Geol 6:45–55

Myers EM, Janzen FJ, Adams DC (2006) Quantitative genetics of plastron shape in slider turtles (*Trachemys scripta*). Evolution 60:563–572

Novack-Gottshall PM, Lanier MA (2008) Scale-dependence of Cope's rule in body size evolution of paleozoic brachiopods. Proc Nat Acad Sci 105:5430–5434

Peck LS (2008) Brachiopods and climate change. Earth Environ Sci Trans Royal Soc Edinb 98:451–456

Pitrat CW, Rogers FS (1978) *Spinocyrtia* and its epizoans in the traverse group (Devonian) of Michigan. J Paleontol 52:1315–1324

Polly PD (2003) Paleophylogeography: the tempo of geographic differentiation in marmots (marmota). J Mammal 84:369–384

Polly PD (2004) On the simulation of the evolution of morphological shape: multivariate shape under selection and drift. Palaeontol Electron 7.2.7A:1–28

Polly PD (2005) Development, geography, and sample size in P matrix evolution: molar-shape change in island populations of Sorex araneus. Evol Dev 7:29–41

Polly PD (2007) Adaptive zones and the pinniped ankle: a 3D quantitative analysis of carnivoran tarsal evolution. In: Sargis E, Dagosto M (eds) Mammalian evolutionary morphology: a tribute to Frederick S. Szalay. Springer, Dordrecht, The Netherlands

Richards RP, Shabica CW (1969) Cylindrical living burrows in Ordovician dalmanellid brachiopod beds. J Paleontol 43:838–841

Richards RP (1972) Autecology of Richmondian brachiopods (late Ordovician of Indiana and Ohio). J Paleontol 46:386–405

Rodland DL, Kowalewski M, Carroll M, Simoes MG (2004) Colonization of a 'Lost World': encrustation patterns in modern subtropical brachiopod assemblages. Palaios 19:381–395

Rodland DL, Kowalewski M, Carroll M, Simoes MG (2006) The temporal resolution of epibiont assemblages: are they ecological snapshots or overexposures? J Geol 114:313–324

Rodrigues SC, Simoes MG, Kowalewski M, Petti MAV, Nonato EF, Martinez S, Rio CDJ (2008) Biotic interaction between spionid polychaetes and bouchardiid brachiopods: paleoecological, taphonomic and evolutionary implications. Acta Palaeontologica Polonica 53:657–668

Rohlf FJ (1990) Morphometrics. Annu Rev Ecol Sys 21:299–316

Rohlf FJ (1998) On applications of geometric morphometrics to studies of ontogeny and phylogeny. Sys Biol 47:147–158

Rook L, O'Higgins P (2005) A comparative study of adult facial morphology and its ontogeny in the fossil macaque Macaca majori from Capo Figari, Sardinia, Italy. Folia Primatologica 76:151–171

Rudwick MJS (1962) Filter-feeding mechanisms in some brachiopods from New Zealand. J Linn Soc London Zool 44:592–615

Rudwick MJS (1970) Living and fossil brachiopods. Hutchinson University, London

Schneider CL, Leighton LR (2007) The influence of spiriferide micro-ornament on Devonian Epizoans. Geol Soc Am Abstr 39:531

Spjeldnaes N (1984) Epifauna as a tool in autecological analysis of silurian brachiopods. Spec Pap Paleontol 32:225–235

Stanley SM (2005) Earth system history. W. H. Freeman and Company, New York

Stayton CT (2006) Testing hypotheses of convergence with multivariate data: morphological and functional convergence among herbivorous lizards. Evolution 60:824–841

Taylor PD, Wilson MA (2002) A new terminology for marine organisms inhabiting hard substrates. Palaios 17:522–525

Taylor PD, Wilson MA (2003) Palaeoecology and evolution of marine hard substrate communities. Earth Sci Rev 62:1–103

Thayer CW (1974) Substrate specificity of Devonian Epizoa. J Paleontol 48:881–894

Thayer CW (1977) Recruitment, growth, and mortality of a living articulate brachiopod, with implications for the interpretation of survivorship curves, pp. 98–109

Tomasovych A, Carlson SJ, Labarbera M (2008) Ontogenetic niche shift in the brachiopod *Terebratalia transversa*: relationship between the loss of rotation ability and allometric growth. Palaeontology 51:1471–1496

Webber AJ, Hunda BR (2007) Quantitatively comparing morphological trends to environment in the fossil record (Cincinnatian series; upper Ordovician). Evolution 61:1455–1465

Webster M, Hughes N (1999) Compaction-related deformation in Cambrian olenelloid trilobites and its implications for fossil morphometry. J Paleontol 73:355–371

Webster M, Zelditch ML (2011) Modularity of a Cambrian ptychoparioid trilobite cranidium. Evol Dev 13:96–109
Zelditch ML, Lundrigan BL, Garland T (2004) Developmental regulation of skull morphology: ontogenetic dynamics of variance. Evol Dev 6:194–206

Chapter 2
Testing the Taxonomy and Phylogeny of Eastern North American Atrypid Brachiopods: A Geometric Morphometric Approach

2.1 Introduction

2.1.1 Taxonomy

All atrypide brachiopods in North America were once referred to a single collective species, *Atrypa reticularis* (e.g., Fenton and Fenton 1930). *Atrypa reticularis sensu stricto* is now recognized only from the Silurian of Gotland (Copper 2004). The group has been radically revised in the last three decades and the understanding of the evolutionary relationships between genera is still in flux. The brachiopods that were once referred to one species are now distributed among 38 genera in five subfamilies within the family Atrypidae (Copper 1973, 1996, 2001a, 2002, 2004; Day 1998; Day and Copper 1998; Williams et al. 2002). Although the taxonomy of Copper's (1973) phylogeny is now partially out of date, having been revised again by Copper (2001a) based on differences in shell size, shell shape, surface ornamentation and internal morphological features, his phylogenetic hypothesis remains the only one for atrypides. To date, no one has attempted to quantify the morphological characters of these genera to test whether shell shape evolution is consistent with the phylogenetic arrangement. In this study, genera from the Atrypinae, Variatrypinae and Spinatrypinae are studied based on external morphological characters.

The genus *Atrypa* has been revised extensively since 1965 and has been split into several genera (Alvarez 2006). *Atrypa* was most closely related to *Desquamatia* and *Spinatrypa* (Boucot 1964) which were then all referred to Atrypinae (Williams and Rowell 1965). The phylogenetic relationships of *Atrypa*, *Gotatrypa*, *Kyrtatrypa* and many other atrypids were later studied by Copper (1973) who suggested that *Atrypa* is more closely related to *Gotatrypa* than to *Spinatrypa* or *Pseudoatrypa*. Based on that phylogeny and subsequent work, *Desquamatia* is today referred to Variatrypinae and *Spinatrypa* is referred to Spinatrypinae (Copper 2002).

R. Bose, *Biodiversity and Evolutionary Ecology of Extinct Organisms*,
Springer Theses, DOI: 10.1007/978-3-642-31721-7_2,
© Springer-Verlag Berlin Heidelberg 2013

Atrypid individuals that have been measured for this study belong to the Atrypinae, Variatrypinae and Spinatrypinae subfamilies, which have similar dorsally to dorso-medially directed spiralia (Copper 1996, 2002), distinct jugal processes, and distinct types of pedicle collar attachments to deltidial plates (Copper 1967, 1977). These three subfamilies were selected for this study as representatives of the family Atrypidae as they represent the longest stratigraphic ranges within the Silurian and Devonian time periods.

Atrypin adults have wavelike, overlapping or imbricate growth lamellae extended as frills, with loss of pedicle opening during ontogeny in most shells; while variatrypins have widely separated growth lamellae extended as expansive frills or alternatively have reduced growth lamellae but with simple tubular ribs, with most forms retaining the pedicle foramen. Spinatrypinae have commonly spinose, short growth lamellae with ribs disrupted into waves, producing nodular surface macro-ornament, with all forms retaining the pedicle foramen (Copper 2001a, 2002).

Copper (1973) used rib structure, pedicle structure, structure of jugal processes and dentition type to construct his phylogeny of four atrypide families. Shell shape was not considered as a criterion in his (1973) phylogenetic reconstruction. In our study, atrypid shell shape was evaluated using geometric morphometrics on nine genera representing three subfamilies of Atrypidae recognized by Copper's taxonomy and phylogeny. Our data from the Early Silurian to Early Devonian samples consist of individuals that belong to the Atrypinae while the Middle Devonian to Late Devonian samples consist of those from the Variatrypinae and Spinatrypinae subfamilies (Table 2.1). We analyzed shell shape to see if the patterns of differentiation are consistent with the taxonomic and phylogenetic structure proposed by Copper. A complete phylogenetic analysis based on internal morphological characters of these diverse atrypid genera in the Atrypinae, Variatrypinae and Spinatrypinae subfamilies from the eastern North American region awaits future analysis. Overall, this study tests whether the differences in shell shape between genera are consistent with Copper's classification and whether the quantitative results support the phylogeny of the atrypid brachiopods.

2.1.2 Hypotheses

Six hypotheses were tested in this study: (1) if the Atrypinae, Variatrypinae and Spinatrypinae are distinct subfamilies, as proposed by Copper (1973) and Williams et al. (2002), then the average morphology of these subfamilies should show significant differences between them; (2) likewise, if the genera within subfamilies are truly distinct, then the average shell shape between these genera should show morphometric differences; (3) furthermore, if genera are correctly referred to subfamilies, then average morphological distance between genera in different subfamilies should be greater than that from within one subfamily; (4) If evolutionary stasis is predominant in the P3 evolutionary ecological unit (EEU, see below) caused by ecological interlocking or other environmental influences, then,

Table 2.1 Geographic locations for atrypid taxa with specific information on time interval and stage, EE subunit, and depositional environment

Nos.	Locality	Sample size	Taxonomy	Formation	Period	Stage	EE Unit (duration in my)	Basin	Environment
18	North-central Iowa, USA	123	*Pseudoatrypa*, Variatrypinae	Cerro Gordo Member, Lime Creek Shale, Hackberry Grove	Devonian (Late)	Frasnian	Genesee (5–6)	Southern margin of the Iowa Basin	Mid-outer carbonate shelf: restricted and open marine oxic facies
17	NE Michigan, USA	210	*Pseudoatrypa*, Variatrypinae	Traverse Group	Devonian (Middle)	Late Eifelian-Middle Late Givetian	Hamilton-Tully (6–7)	Flanks of the northeastern part of Michigan Basin	Mixed setting (siliciclastics and carbonates)
16	Western New York, USA	134 23	*Pseudoatrypa*, Variatrypinae *Spinatrypa*, Spinatrypinae	Hamilton Group	Devonian (Middle)	Late Eifelian-Middle Late Givetian	Hamilton-Tully (6–7)	Folded ridge valley province of northern Appalachian Basin	Siliciclastics bordered by carbonate shelf
15	Southeastern Indiana, USA	63	*Pseudoatrypa*, Variatrypinae	Hamilton Silver Creek member of the North Vernon limestone	Devonian (Middle)	Late Eifelian-Middle Late Givetian	Hamilton-Tully (6–7)	Cincinnatian Arch between Illinois and Michigan Basin	Carbonate
14	Fulton, Central Missouri, USA	33	*Pseudoatrypa*, Variatrypinae	Callaway Limestone	Devonian (Middle)	Late Eifelian-Middle Late Givetian	Hamilton-Tully (6–7)	Southern Iowa Basin	Carbonate
13	Ohio, USA	9	*Pseudoatrypa*, Variatrypinae	Dundee Limestone	Devonian (Middle)	Lower-Mid Eifelian	Onondaga (5–6)	Northwestern flank of the Findlay Arch	Carbonate
12	Central New York, USA	34	*Spinatrypa*, Spinatrypinae	Onondaga Limestone	Devonian (Middle)	Lower-Mid Eifelian	Onondaga (5–6)	Ridge and valley province of central Appalachian Basin	Carbonate

Table 2.1 (continued)

Nos.	Locality	Sample size	Taxonomy	Formation	Period	Stage	EE Unit (duration in my)	Basin	Environment
11	Eastern Tennessee, USA	151	*Atrypa*, Atrypinae	Birdsong Shale?, Linden Group, Helderbergian	Devonian (Early)	Emsian	Schoharie (5)	On the east along flanks of the Nashville Dome	Siliciclastics
10	Central-east New York, USA	33 1	*Atrypa*, Atrypinae *Kyrtatrypa*, Atrypinae	Schoharie Grit, New Scotland Limestone of Helderbergian Group	Devonian (Early)	Emsian	Schoharie (5)	Northeast-trending folds in the Valley and Ridge Province of the central Appalachian Basin	Shallow carbonates and siliciclastics.
9	West Virginia, USA	46 4	*Atrypa*, Atrypinae *Kyrtatrypa*, Atrypinae	Keyser Limestone	Devonian (Early)	Lockhovian	Helderberg (6)	On a carbonate ramp in the valley and ridge province of the central Appalachian Basin	Carbonate
8	Maryland, USA	11 17	*Atrypa*, Atrypinae *Kyrtatrypa*, Atrypinae	Keyser Limestone	Devonian (Early)	Lockhovian	Helderberg (6)	On a carbonate ramp in the valley and ridge province of the central Appalachian Basin	Carbonate
7	Central-east New York, USA	34 28	*Atrypa*, Atrypinae *Kyrtatrypa*, Atrypinae	Lower Helderberg Group	Devonian (Early)	Lockhovian	Helderberg (6)	Northeast-trending folds in the Valley and Ridge Province of the central Appalachian Basin	Carbonate
6	Oklahoma, USA	76 13 2	*Atrypa*, Atrypinae *Kyrtatrypa*, Atrypinae *Endrea*, Atrypinae	Hunton (Haragan) Limestone (Yellow shale below limestone)	Devonian (Early)	Lockhovian	Helderberg (6)	'Hunton Ramp' on the margin of the Oklahoma Aulacogen and Ouachita Trough	Carbonate ramp

Table 2.1 (continued)

Nos.	Locality	Sample size	Taxonomy	Formation	Period	Stage	EE Unit (duration in my)	Basin	Environment
5	Oklahoma, USA	79 10 3	*Atrypa,* Atrypinae *Gotatrypa,* Atrypinae *Oglupes?,* Atrypinae	Henryhouse Limestone, Hunton Group	Silurian (Late)	Ludlow	Keyser (2)	Surface of Anadarko Basin	Carbonate
4	West-Central Tennessee, USA	64 81 5	*Atrypa,* Atrypinae *Gotatrypa,* Atrypinae *Endrea,* Atrypinae	Waldron Clay member, Wayne Formation	Silurian (Middle)	Wenlock	Upper Clinton-Lockport (7–8)	Northern and western margins of the Central Basin of Tennessee	Siliciclastics in the thick limestone sequence
3	Indiana, USA	18 43 8	*Atrypa,* Atrypinae *Endrea,* Atrypinae *Oglupes?,* Atrypinae	Niagara Group	Silurian (Middle)	Wenlock	Upper Clinton-Lockport (7–8)	Cincinnatian Arch between Illinois and Michigan Basin	Carbonates
2	Western New York, USA	41 54 3 1	*Atrypa,* Atrypinae *Gotatrypa,* Atrypinae *Endrea,* Atrypinae *Oglupes?,* Atrypinae	Lockport Formation	Silurian (Middle)	Wenlock	Upper Clinton-Lockport (7–8)	Thrusts along ridge and valleys of central Appalachian Basin	Siliciclastics
1	Quebec, Canada	2 111 18 7	*Atrypa,* Atrypinae *Gotatrypa,* Atrypinae *Joviatrypa,* Atrypinae *Dihelictera,* Atrypinae	Jupiter Formation	Silurian (Early)	Llandovery	Lower Clinton (4)	Anticosti Basin	Shallow water shelly packstones to middle-outer shelf micritic mudstones

despite taxonomic replacement, morphology is expected to remain the same with substantial overlap between genera through time; (5) If evolutionary stasis is predominant in the P3 EEU, then minor morphological change is expected within each genus through time; (6) If biogeographic differences in shell shape among geographic locations are due to provinciality, then at a given time, genera from the same paleogeographic locations are expected to cluster and morphological distances among paleogeographic locations are expected to be similar to those observed between genera.

2.1.3 Ecological Evolutionary Units and Subunits

The Silurian and Devonian have been classified into a single Paleozoic Ecological Evolutionary Units (EEU) (designated as P3), which was a time marked by periods of stability (designated as subunits or Ecological Evolutionary Subunits (EESUs)) (Boucot 1983, 1986, 1990a, b, c; Sheehan 1991, 1996; Holterhoff 1996; Brett et al. 2009) interspersed by periods of minor reorganization and extinction (Brett et al. 1990; Brett and Baird 1995; Holterhoff 1996; Sheehan 1996; Ivany et al. 2009). The major extinction events of the end-Ordovician and Late Devonian mark the P3 EEU boundaries (Sheehan 1996). The eight P3 EESUs included in the present study are shown in Fig. 2.1, Table 2.1. The P3 EESUs have been well studied (Brett and Baird 1995; Brett et al. 2009) in the Appalachian Basin with respect to community stability patterns, but studies involving morphological shape change or stasis within brachiopod species lineages from these EESUs are lacking excepting Lieberman et al. (1995) who studied two brachiopod species lineages for stasis. In general, brachiopods were abundant, diverse and well-preserved during this time interval, providing plentiful data for morphological shape study. Morphological shape change patterns in P3 EEU atrypid subfamilies and genera are described here to trace both temporal and spatial variation within these brachiopod genera. Thus, this study is designed to determine comparative morphological shape patterns within the atrypid brachiopods belonging to the Atrypinae, Variatrypinae and Spinatrypinae from the strata making up the P3 EEU from the Appalachian Basin and their stratigraphic equivalents within the Eastern American biogeographic Realm, spanning the entire 64 Myr of the Silurian-Devonian (441–376 Myr) rock record.

2.2 Materials

2.2.1 Paleogeography

During the Silurian-Devonian time, the Eastern American Realm was relatively isolated from other biogeographic realms and was indeed a realm of its own. Though these realms had established connections between them in this long interval of time during major transgressive events of the transgressive-regressive sea level cycles, much of the evolution of these atrypide lineages possibly occurred within the basin. Boucot (1975) and Boucot and Blodgett (2001) referred to the Eastern American Realm as a warm or hot unit with lower to mid latitudinal strata rich in evaporites, redbeds, carbonate rocks and reef developments. Thus, Silurian and Devonian of Eastern North America, representing high taxic diversity at all taxonomic levels, make an important biogeographic realm for taxonomic investigation.

The Silurian was a period of marked provincialism for the brachiopod faunas (Boucot and Blodgett 2001) during which the Acadian orogeny occurred (Van der Pluijm 1993) and shallow marine carbonate deposition was widespread (Berry and Boucot 1970). During the Llandovery (early Silurian), sea levels were low and a comparatively cool climate was indicative from the less extensive reef deposits (Copper 2001b) and thus, evolution of some endemic atrypide genera were possibly restricted in their small environmental regimes. Climates warmed up and sea levels started rising during the Wenlock (middle Silurian) as evidenced by the onset of reef growth (Copper 1973, 2004). Abundant patch reefs occurred in parts of Michigan, Ontario, Ohio and Indiana (Cumings and Shrock 1928; Lowenstam 1957). A shallow marine environmental setting is suggested by the presence of mid-platform carbonates in eastern North America during this time. Thus, our samples (*Atrypa, Gotatrypa, Endrea*) from the Middle Silurian of Appalachian basin, Cincinnatian Arch and central Tennessee basins were somewhat similar during this time. Some of the Early Silurian genera (*Joviatrypa, Dihelictera*) remained restricted to the Hudson Bay lowlands. Some genera (*Atrypa, Gotatrypa*) in Anadarko basin of Oklahoma localities still persisted in the carbonate platforms during the Ludlow time. By the Pridoli time (late Silurian), sea level dropped leading to evaporitic conditions in some basins and closure of many sea connections between basins, including between eastern North America and Europe. This provincialization continued during much of the Early Devonian (Copper 1973) of Eastern North America, which gave rise to some endemic genera (*Kyrtatrypa*) in the margins of Oklahoma aulacogen, central Appalachian basin (central-east New York, West Virginia and Maryland) and the eastern Tennessee Nashville Dome localities with continued persistence of the *Atrypa* lineage in these localities. During the Early Devonian and early Middle Devonian (Lockhovian-early Eifelian), sea level was low and most genera (*Atrypa, Kyrtatrypa*) from the Appalachian, Michigan, Iowa and Anadarko basins were most likely separated by geographic barriers (Findley, Kankakee and Cincinnatian Arches) which persisted through the Middle Devonian, thus, giving rise to new evolutionary lineages (*Pseudoatrypa, Spinatrypa*) in some early Eifelian localities (Ohio and central New York). All of these intracratonic arches served as barriers to shallow marine dispersal during the Devonian (Koch and Day 1995; Rode and Lieberman 2005). Later in the Middle Devonian (late Eifelian to early Givetian), sea levels had risen again and climates warmed up with widespread carbonate deposition in this region. This marked sea-level rise during this time possibly breached the Ozark dome, Wisconsin, Findley, Kankakee, Cincinnatian Arches and Acadian

Highlands of Eastern North America further facilitating mixing of faunas within all geographic localities in the Eastern American biogeographic realm. The late Givetian, or early Frasnian (Late Devonian) was thus, a marked period of cosmopolitanism when Appalachian sea lanes were connected all over again and thus similar genera (*Pseudoatrypa*) persisted across various geographic localities (Michigan Basin, Iowa and Missouri localities in Iowa Basin, northern Appalachian Basin and Cincinnatian Arch) during this time. At the end of the Frasnian time, black shales or disconformities were produced locally with sea levels continuing to rise. Eventually, the muddy bottom dwelling and stenohaline atrypides went extinct at the end of the Frasnian more likely due to ecological replacement of these faunas by other higher order organisms (Copper 1973). Thus, this varied biogeographic setting makes it all the more interesting to investigate the taxonomic composition in atrypides during the Silurian-Devonian time period in eastern North America.

Of the eighteen sampled geographic localities in this study, samples from Hudson Bay lowlands represent inner shelf environments, those from Appalachian (New York, Maryland, West Virginia) localities represent inner to middle shelf environments, Tennessee localities represent inner shelf environment with some derived clastics in the north, Michigan locality represents middle shelf environment, Cincinnatian Arch (Indiana) locality represents middle shelf environment, Ohio locality represent environments dominated by eastern derived clastics, Missouri locality represent inner shelf environments, Iowa locality represents inner shelf environments with presence of evaporite beds, and Oklahoma localities represent environments that ranged from inner to middle shelves (Day 1998).

Thus, for this study, geographic variation was examined in a few genera from the Middle Silurian, Early Devonian and Middle Devonian Eastern North American localities. The stratigraphic, lithologic and paleogeographic settings for the sampled atrypids from Silurian-Devonian of Eastern North America in this study are given in Table 2.1.

2.3 Methods

2.3.1 Data Set

We tested morphological variation using a total of 1593 dorsal and ventral valves (Table 2.1) of well preserved atrypid brachiopods. Of those specimens, 1300 specimens were used to test morphological evolution within *Atrypa*, *Gotatrypa* and *Pseudoatrypa*; and 904 specimens were used to assess geographic variation within *Atrypa* from the Middle Silurian and Early Devonian localities, *Kyrtatrypa* from the Early Devonian localities, and *Pseudoatrypa* from the Middle Devonian localities in eastern North America. Specimens were identified to genus level and grouped within their respective subfamilies (Atrypinae = 964, Variatrypinae = 572, and Spinatrypinae = 57). The geographic location and respective

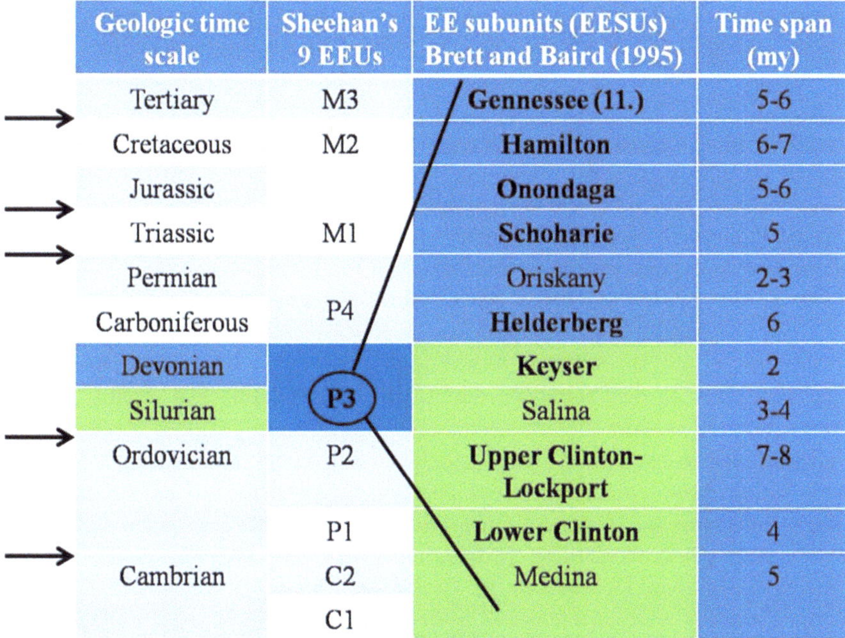

Geologic time scale	Sheehan's 9 EEUs	EE subunits (EESUs) Brett and Baird (1995)	Time span (my)
Tertiary	M3	**Gennessee (11.)**	5-6
Cretaceous	M2	**Hamilton**	6-7
Jurassic		**Onondaga**	5-6
Triassic	M1	**Schoharie**	5
Permian		Oriskany	2-3
Carboniferous	P4	**Helderberg**	6
Devonian		**Keyser**	2
Silurian	P3	Salina	3-4
Ordovician	P2	**Upper Clinton-Lockport**	7-8
	P1	**Lower Clinton**	4
Cambrian	C2	Medina	5
	C1		

Fig. 2.1 Ecological evolutionary unit P3 (*circled*) showing the major subdivided 11 EE subunits in the Silurian and Devonian (data sampled from the 8 EESUs are marked in *bold*). (after Brett and Baird 1995)

sample sizes are reported on the map in Fig. 2.2 and in Table 2.1. All specimens were identified based on external morphological characters and ornamentation (Fig. 2.3). The material we used is housed in the Invertebrate Paleontology Collections of the American Museum of Natural History, Yale Peabody Museum, New York State Museum and Indiana University Paleontology Collections.

2.3.2 Geometric Morphometrics

Geometric morphometrics is the analysis of geometric landmark coordinate points on specific parts of an organism (Bookstein 1991; MacLeod 2002; Zelditch et al. 2004; Webster 2011). Morphometric analysis is based on the use of landmarks to capture shape (Rohlf and Marcus 1993); landmarks are points representing the same location on each specimen. In this study, we used 9 two-dimensional landmark points to capture the most meaningful shape differences (Fig. 2.4). Landmarks were digitized from image files using Thin Plate Spline Dig software (Rohlf 2004). When selecting landmarks for analyses, we chose points that not only characterized body shape accurately, but also represented some aspect of the inferred ecological niche. These landmarks represent discrete points that correspond among forms (sensu Bookstein 1991) and are appropriate for analyses attempting to capture shape changes or function. These points are at the intersection of articulation of both valves except landmarks 1 and 9 (1 = umbo tip on dorsal valve on the plane of symmetry; 2 and 8 = left and right posterior marginal tips of the hingeline; also region for food intake from inhalant currents; 3 and 7 = mid shell tips along the widest region of the shell; 4 and 6 = anterior commissure marginal ends; 5 = anterior margin of commissure on the plane of symmetry; 9 = beak tip on ventral valve on the plane of symmetry). The same eight landmarks (1–8 on dorsal valves and 2–9 on ventral valves) were used to compare both dorsal and ventral valves (Fig. 2.4).

Procrustes analysis (Rohlf 1990; Rohlf and Slice 1990; Rohlf 1999; Slice 2001) was performed on original shape data, rotating, translating and scaling all landmarks to remove size effects while maintaining their geometric relationships. Pairwise Procrustes distances were calculated between the mean shapes of genera both within Atrypinae and between Atrypinae, Variatrypinae and Spinatrypinae (Fig. 2.5). Procrustes distances were also calculated between mean shell shapes within each genus from different time units and between mean shell shapes of different geographic localities in the Middle Silurian, Early and Middle Devonian. These distances were all measured in Procrustes units. Procrustes units are measures of shape difference in multivariate space, whose units are arbitrarily derived from the landmark data, but they are comparable across objects with the same number of landmarks (Rohlf 1990; Rohlf and Slice 1990). Principal component analysis was performed on the covariance matrix of Procrustes residuals to determine the morphological variation between the Atrypinae, Variatrypinae and Spinatrypinae and among genera within Atrypinae. Principal component analysis was also performed to determine within genus variation in time and space units.

2.3.3 Morphometric Divergence

Evolutionary rate and mode in morphological divergence were assessed using the maximum-likelihood method of Polly (2008). This method estimates the mean per-step evolutionary rate and the degree of stabilizing or diversifying selection from a matrix of pairwise morphological distances and divergence times. Morphological distance was calculated as pairwise Procrustes distances among genera (Fig. 2.5) and divergence time was calculated using the patristic distance in millions of years on phylogenetic tree of Copper (1973) (Fig. 2.6), which is an estimate of the total time in millions of years that the two genera have been diverging independently since they last shared a common ancestor. The method uses the following equation to estimate rate and mode simultaneously,

$$D = rt \,^\wedge\, a, \tag{2.1}$$

where D is morphological divergence (procrustes distance), r is the mean rate of morphological divergence, t is divergence time, and a is a coefficient that ranges from 0 to 1, where 0 represents complete stabilizing selection (stasis), 0.5 represents perfect random divergence (brownian motion) and 1 represents perfect diversifying (directional) selection (Polly 2008). Maximum-likelihood is used to find the parameters r and a that maximize the likelihood of the data, and are thus the best estimates for rate and mode. The data were bootstrapped 1000 times to generate standard errors for these estimates. This method is derived directly from the work presented by Polly (2004) and is mathematically related to other methods in evolutionary genetics (Lande 1976; Felsenstein 1988; Gingerich 1993; Roopnarine 2003; Hunt 2007).

Lastly, a few atrypid genera (*Atrypa*, *Gotatrypa*, *Endrea*, *Pseudoatrypa*) were tested for morphological shape differences in specimens from shale and carbonate lithologic settings. For example, *Atrypa* specimens were tested for differences between Lower Devonian Linden Group of Tennessee (siliciclastic) and Keyser Limestone of Maryland (carbonate). Similar tests were performed for other genera.

2.3.4 Statistical Analysis

We performed several statistical tests to assess morphological distinctness between atrypin genera, between atrypid subfamilies and to investigate the correctness of the phylogenetic relatedness between these genera. Multivariate analysis of variance (MANOVA) was performed to test for significant morphological shape difference (a) between three subfamilies, (b) between genera within one subfamily, (c) within genera between time and space units, and (d) within genera between shale and carbonate environments. Discriminant function analysis (DFA) was also performed to highlight the morphological differentiation within and between subfamilies. Pairwise distance between genera from within a subfamily was then compared with those between subfamilies. A bootstrap test was performed to draw statistical inference regarding the frequency (out of 1000 iterations) of randomly observing the difference in mean sample morphology between time units. The trend in mean shape through time was constructed for individual genera from Principal component scores. Euclidean cluster analysis (UPGMA) was performed to identify similarities in individual genera sampled from different time intervals, and from geographic intervals at a given time. Average Euclidean distances in time and space were also compared to assess whether the temporal distances are similar to what one would expect from the replacement of one population by a geographically distinct one (Polly 2003).

Fig. 2.2 Sampled localities for atrypid brachiopods in eastern North America. Filled triangle in black indicates Silurian localities. Filled square in black indicates Devonian localities. Numbers 1–18 indicate 18 different localities from where samples were collected. For detailed locality information see Table 2.1

2.4 Results

2.4.1 Taxonomic Differentiation

Based on qualitative phenotypic traits, the specimens were identified to genus level (Fig. 2.3, Table 2.1). Atrypin genera were distinguished from other genera by their characteristic closely spaced growth lamellae and loss of the pedicle opening. The most distinguishable phenotypic characters of the variatrypin genera included the long, tubular rib structure, pedicle opening and wide spacing between growth lamellae; while spinatrypin genera possessed spinose growth lamellae with highly imbricated tubular rib structure. Frills were not preserved in most atrypides, so it could not be utilized in distinguishing the atrypid members in various subfamilies. Our sample includes the following genera: Atrypinae—*Atrypa, Gotatrypa, Endrea,*

Fig. 2.3 Dorsal views of genera from Atrypinae, Variatrypinae, and Spinatrypinae subfamilies: **a** YPM 224604, *Atrypa*; **b** YPM 224444, *Gotatrypa*; **c** NYSM E2341 62-5, *Kyrtatrypa*; **d** YPM 224240, *Joviatrypa*; **e** YPM 224522, *Endrea*; **f** YPM 224450, *Dihelictera*; **g** YPM 225957, *Spinatrypa*; **h** YPM 226001, *Spinatrypa*; **i** YPM 226006, *Spinatrypa* (note the spinose imbricated lamellae in lower right area of the shell in **h**, and widened spaces between growth lamellae in **i**; **j** YPM 225921, *Pseudoatrypa*. Note: the scale bar is the *upper vertical line* 1 cm for specimens **a–f** and the *lower horizontal line* 1 cm for specimens **g–j**

Joviatrypa, Kyrtatrypa, Dihelictera, Oglupes?, Protatrypa, and *Rugosatrypa*; Variatrypinae—*Pseudoatrypa,* and *Desquamatia (Independatrypa)*; Spinatrypinae—*Spinatrypa*. Note that *Protatrypa, Rugosatrypa* and *Desquamatia (Independatrypa)* were not included in the morphometric analysis because of their small sample sizes.

Principal component analysis showed significant overlap between the subfamilies and genera. MANOVA found significant differences between mean shell shape in the three subfamilies for dorsal valves (F = 30.7, df1 = 24, df2 = 3130, $p < 0.01$) (Fig. 2.4b). MANOVA (post hoc pairwise tests with Bonferroni correction) also found significant differences among mean shape in the following genera within Atrypinae for dorsal valves (F = 4.389, df1 = 72, df2 = 5060, $p < 0.01$) and ventral valves (F = 3.628, df1 = 72, df2 = 5027, $p < 0.01$) (Fig. 2.4c, Table 2.2): *Atrypa-Gotatrypa, Atrypa-Endrea, Atrypa-Kyrtatrypa, Endrea-Kyrtatrypa, Gotatrypa-Kyrtatrypa, Kyrtatrypa-Joviatrypa,* and *Oglupes?-Gotatrypa*. Significant differences in mean shape were also found between

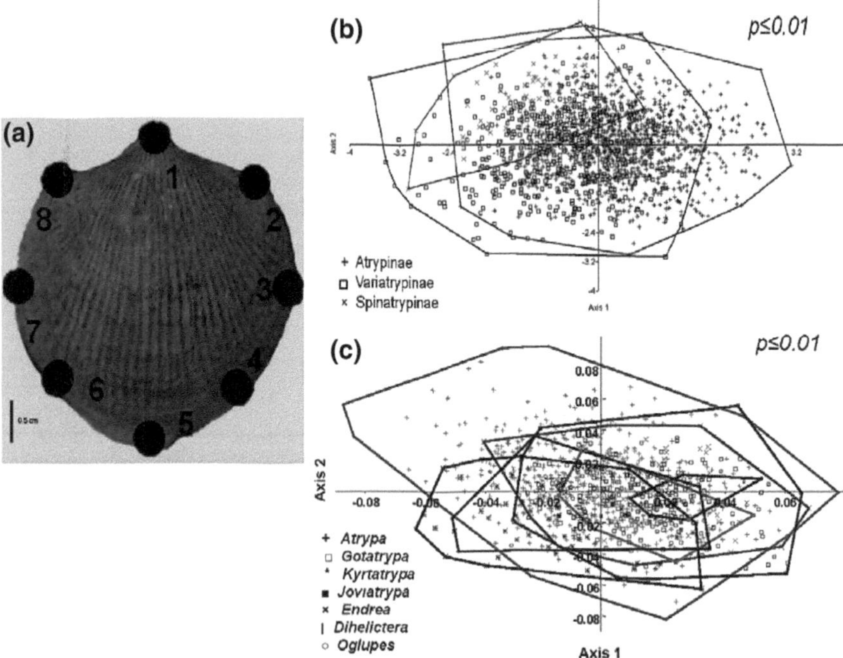

Fig. 2.4 **a** Location of eight landmarks on the pedicle valve of an atrypid sample for geometric morphometric analysis; **b** CVA plot showing morphometric differences between Atrypinae, Variatrypinae, and Spinatrypinae subfamilies (*p < 0.01*); **c** CVA plot showing morphological differentiation between genera within Atrypinae subfamily (*p < 0.01*). Note that *Rugosatrypa* and *Protatrypa* have been removed from analysis as these were only one member from each genus

Pseudoatrypa-Spinatrypa, from Variatrypinae and Spinatrypinae respectively (*p < 0.01*). On average, the mean shape difference between subfamilies ranged from 0.1 to 1.3 Procrustes units, and the difference between genera within the Atrypinae subfamily ranged from 0.01–0.05 Procrustes units (Fig. 2.5).

Maximum-likelihood estimation of the rate and mode of evolution given phylogeny of (Copper 1973) (Fig. 2.6) yielded a rate of 0.012 ± 0.12 Procrustes units per million years and a mode coefficient a of 0.97 ± 0.15, indicating that diversifying selection has made the means of these atrypid genera more different than one would expect by random evolution (Fig. 2.7). In random evolution or Brownian motion, the direction and intensity of selection would have caused change in morphology over time but with changeable conditions (Polly 2004). A general observation suggests that morphometric shape differences between pairs of genera increases with phylogenetic distance with some discrepancy in *Pseudoatrypa* (Variatrypinae) (Table 2.3), which appears to be morphologically closer to the atrypins, more similar to what one would expect for genus-level than sub-family level differentiation.

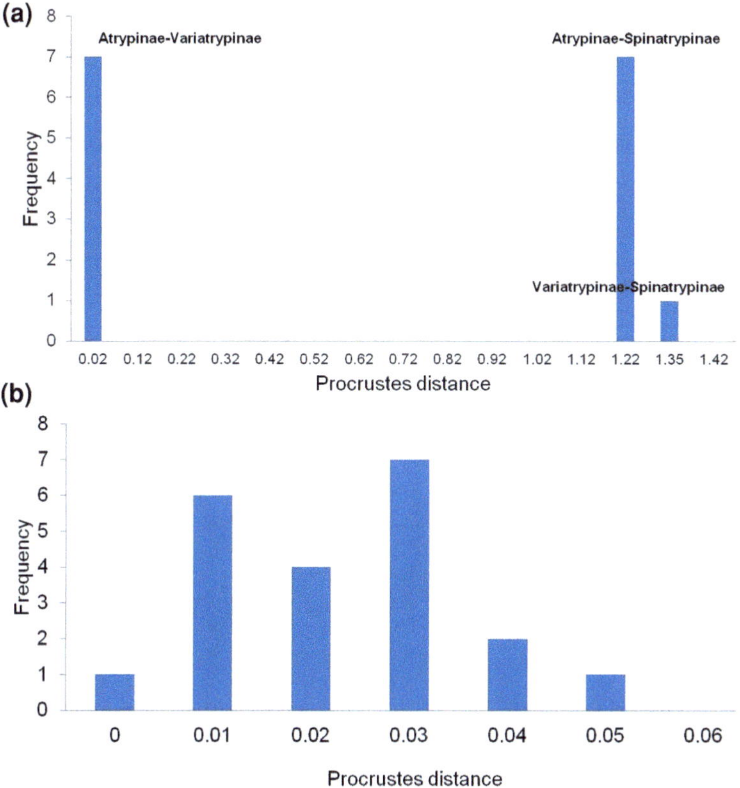

Fig. 2.5 **a** Histogram for Procrustes distance between genera from three subfamilies. All genera from Atrypinae nearly maintain a small distance with Variatrypinae (0.12) while a large distance with Spinatrypinae (1.22–1.25). A large procrustes distance (1.345) between Variatrypinae and Spinatrypinae; **b** Small procrustes distance between genera within Atrypinae subfamily (0.01–0.05)

2.4.2 Temporal Variation

Principal component analysis of atrypid individuals shows morphological variation within each group with considerable morphological overlap among the six clustered groups of atrypids based on six coarse scale time units (Fig. 2.8). MANOVA found significant shape differences in dorsal valves for individual genus between different time horizons (*Atrypa*: F = 4.09, df1 = 24, df2 = 1008, *p < 0.01*; *Gotatrypa*: F = 4.475, df1 = 24, df2 = 484, *p < 0.01*; and *Pseudoatrypa*: F = 10.43, df1 = 12, df2 = 510, *p < 0.01*), suggesting short term changes within a lineage. On average, the difference between Early, Middle and Late Silurian time units in *Gotatrypa* shell shape ranges from 1.1 to 1.8 Procrustes units, the difference between Middle Silurian, Late Silurian and Early Devonian

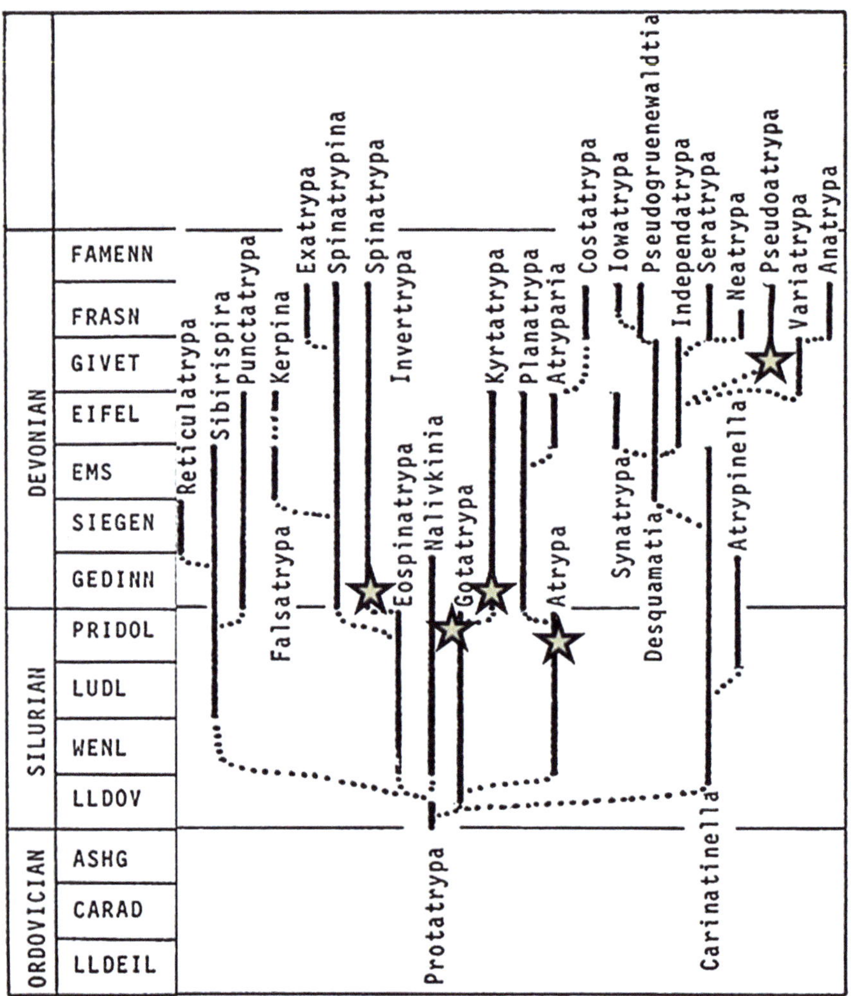

Fig. 2.6 Copper's phylogenetic chart from his 1973 paper that was used to calculate evolutionary divergence times (m.y.) between atrypid genera. The five genera used in this study are marked in the figure as star symbols

time units in *Atrypa* shell shape ranged from 0.7 to 1.1 Procrustes units, and the difference between Middle and Late Devonian time units in *Pseudoatrypa* shell shape was 1.0 Procrustes units (Table 2.4). Overall, these distances were larger than those observed between genera within a subfamily and were comparable to those observed between genera from distinct subfamilies.

2.4.3 Spatial Variation

MANOVA indicated significant geographic shell shape differences among mean shape in individual genera (Middle Silurian *Atrypa*: F = 10.48, df1 = 24, df2 = 214, $p < 0.01$; Early Devonian *Atrypa*: F = 4.18, df1 = 48, df2 = 1165, $p < 0.01$; Early Devonian *Kyrtatrypa*: F = 2.109, df1 = 36, df2 = 136.6, $p < 0.01$; and Middle Devonian *Pseudoatrypa*: F = 5.191, df1 = 48, df2 = 1481, $p < 0.01$).

Dendograms illustrated in Fig. 2.9 depict the similarity in mean valve shape between different geographic localities from the eastern North America region during the three time intervals sampled. Valve morphological shape in the Middle Silurian *Atrypa* shells shows a greater similarity between Tennessee and New York than either region with Indiana (Table 2.5, Fig. 2.9).

During the Early Devonian, *Atrypa* shells from Tennessee and Oklahoma form a close cluster with less morphological distance to the Maryland sample than to the New York and West Virginia samples, which form a cluster with almost similar distance with Maryland (Table 2.5, Fig. 2.9). During the Early Devonian, *Kyrtatrypa* shells from Maryland and New York form a closer cluster with Oklahoma than with West Virginia (Table 2.5, Fig. 2.9). During the Middle Devonian, Missouri samples are more closely linked to those from Michigan and New York than with those from Indiana and Ohio (Table 2.5, Fig. 2.9). Thus, only the Early Devonian *Atrypa* shells show some biogeographic signal.

On average, the shape difference between Middle Silurian *Atrypa* from different geographic regions ranged from 1.7 to 2.3 Procrustes units and the difference between *Atrypa* from different regions in the Early Devonian ranged from 1.0 to 2.0 Procrustes units. The shape difference between *Kyrtatrypa* from different regions in the Early Devonian ranged from 1.4 to 2.8 Procrustes units. The shape difference between *Pseudoatrypa* samples from different Middle Devonian geographic regions ranged from 0.6 to 2.3 Procrustes units. Overall, geographic variation within genera is greater than temporal variation.

Average Procrustes distance over time suggests that the magnitude of morphological shape change was similar in the dorsal and ventral valves (0.025 and 0.028 Procrustes units respectively). Likewise, average Procrustes distance was similar in the two valves between geographic regions (0.030 and 0.029 for Middle Silurian, 0.027 and 0.029 for Early Devonian, and 0.040 and 0.037 Procrustes units for Middle Devonian respectively). However, geographic variation was slightly greater than temporal variation within these samples as suggested from the range of Procrustes distances within these units (time: 0.025–0.028; space: 0.027–0.040). Geographic variation (0.6–2.3) was also greater than temporal variation (0.7–1.8) when tested for individual genera (Tables 2.4, 2.5).

Lastly, significant statistical differences were observed within each genus between shale and carbonate environments ($p < 0.01$).

Table 2.2 'p' values show distinctness between the genera within Atrypinae subfamily[a]

	Atrypa	Dihelictera	Endrea	Gotatrypa	Joviatrypa	Kyrtatrypa	Oglupes
Atrypa	0	0.262	***0.000***	***0.000***	0.088	***0.000***	0.184
Dihelictera	1	0	0.616	0.685	0.950	0.031	0.780
Endrea	***0.000***	1	0	0.084	0.827	***0.000***	0.242
Gotatrypa	***0.000***	1	1	0	0.655	***0.000***	0.011
Joviatrypa	1	1	1	1	0	***0.004***	0.444
Kyrtatrypa	***0.000***	0.647	***0.000***	***0.000***	0.076	0	0.658
Oglupes	1	1	1	0.229	1	1	0

[a] $p < 0.01$

2.5 Discussion

2.5.1 Taxonomy

In this study, various atrypid genera from the Silurian-Devonian atrypid subfamilies were identified based on phenotypic characters other than shell shape. Overall, significant morphometric differences exist in the shells between subfamilies, thus supporting suggestions put forward by Copper (1973) that these groups are distinct. Significant morphometric differences also exist in the shells between genera within Atrypinae subfamily, also supporting his (Copper 1973) suggestion that they are distinct. However, the differences in external shell morphology and shell shape among subfamilies and genera are small, further evidenced from smaller morphological distances, with considerable shape overlap at all levels. Thus, this suggests that genera would be difficult to distinguish based on shell shape alone. Overall, morphological distance is greater between subfamilies than between genera from the same subfamily, thereby supporting phylogenetic patterns and taxonomic differentiation proposed by Copper (1973).

Though morphometric differences are statistically significant between the shape means of subfamilies and genera, their overall morphological variation overlaps at all levels. This study determined the morphological distances between genera from atrypid subfamilies under investigation and evaluated its relationship with the evolutionary divergence time intervals between genera worked out from the phylogenetic chart proposed by Copper in 1973. His (Copper 1973) phylogeny was the most recent study performed in terms of understanding the generic relationships in atrypides, and so this was used in testing the taxonomy and phylogeny of a few atrypid genera of interest in terms of morphological shape.

The results of the analysis of evolutionary rate and mode indicate that diversifying selection has probably been acting on these atrypid genera, despite the very small morphological divergences among them (Fig. 2.7). Based on the principal component plots (Fig. 2.4), the large degree of morphometric overlap among genera might be interpreted to represent stasis, since none of the atrypid genera have unambiguously diverged from each other. However, the statistical definition

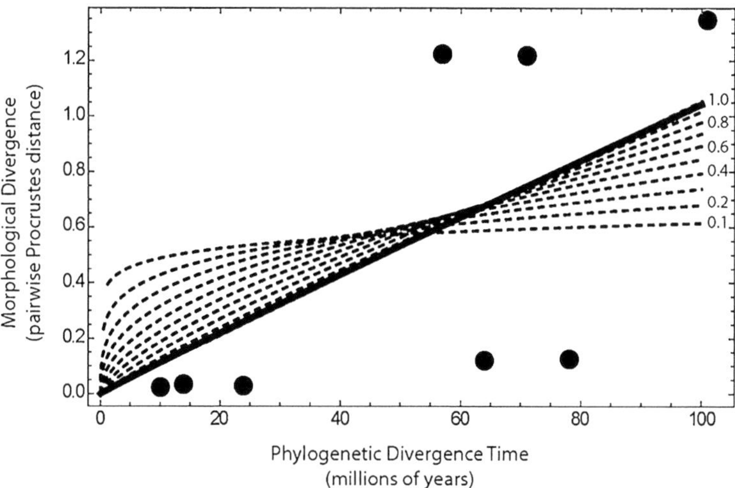

Fig. 2.7 Graph showing morphometric divergence (pairwise Procrustes distances) and phylogenetic divergence (millions of years). The series of *dashed lines* show the expected relationship between morphological and phylogenetic divergence time from strong stabilizing selection (0.1), through random divergence (0.5), to diversifying (directional) selection (1.0). The maximum-likelihood estimate of this relationship, shown by the *dark line*, suggests that these atrypids have experienced diversifying selection

Table 2.3 Procrustes distances between genera within three subfamilies (Atrypinae: *Atrypa, Gotatrypa, Kyrtatrypa,* Variatrypinae: *Pseudoatrypa* and Spinatrypinae: *Spinatrypa*) and evolutionary time between genera calculated from Copper's (1973) phylogenetic chart

Atrypide taxa	Phylogenetic distance (m.y.)	Morphological distance (Procrustes units)
Atrypa-Gotatrypa	10	0.023
Atrypa-Kyrtatrypa	24	0.026
Gotatrypa-Kyrtatrypa	14	0.032
Atrypa-Pseudoatrypa	64	0.121
Gotatrypa-Pseudoatrypa	64	0.122
Kyrtatrypa-Pseudoatrypa	78	0.127
Atrypa-Spinatrypa	57	1.223
Gotatrypa-Spinatrypa	57	1.224
Kyrtatrypa-Spinatrypa	71	1.22
Pseudoatrypa-Spinatrypa	101	1.345

of stasis, or stabilizing selection, is that less divergence has occurred than expected under a random-walk (Brownian motion) model of evolution given the amount of time since divergence and the degree of within-taxon variation (Bookstein 1987; Gingerich 1993; Roopnarine 2001). For these atrypid genera, the changes in mean shape are greater than expected given time since divergence and the amount of within-genus variation—the most likely estimate of a in Eq. 2.1 given the data

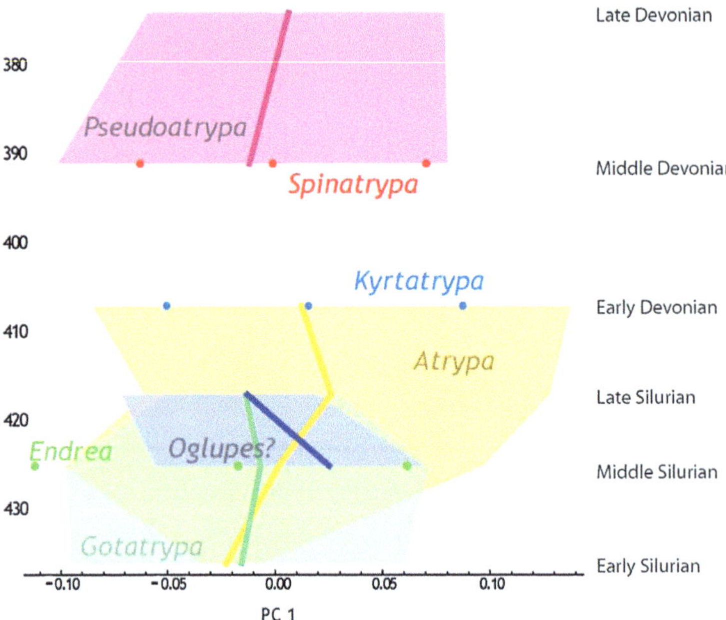

Fig. 2.8 Morphological shape trend for dorsal valves with a minimum, mean and maximum PC scores for seven atrypid genera (*Atrypa, Gotatrypa, Endrea, Oglupes?, Kyrtatrypa, Pseudoatrypa, Spinatrypa*) distributed in the six time units (early Silurian, middle Silurian, Late Silurian, early Devonian, middle Devonian, late Devonian)

presented in Fig. 2.7 is near 1.0 (dark line). Stasis would produce a pattern where the best fit would have a value near 0.0 for parameter *a*, which is decidedly not the case for these data, even when bootstrapped to account for the small sample size and seemingly outlying data points. The best interpretation of shape evolution in these genera given the data is that they were diversifying from one another, but at a rate slow enough that they still overlapped considerably through the time period covered in our study. Our morphometric data are thus consistent with the divisions of taxonomy and the broad strokes of phylogenetic arrangement proposed by Copper (1973), but they indicate that divergences among these genera are very small compared to their within-genus variation, so much so that it is impossible to refer single individuals to a genus on the basis of their geometric shell shape alone. Overall, the scaling between morphological distance and phylogenetic interval generally supports his (Copper 1973) phylogenetic arrangement.

Some general observations noted for a few genera that were included for the test of evolutionary rates and modes are also described. Morphological relatedness between *Pseudoatrypa* and other genera (*Atrypa, Kyrtatrypa, Gotatrypa*) from Atrypinae shows a discrepancy with the evolutionary divergence time proposed by Copper (1973). The lesser morphological shape distance between atrypin and variatrypin members retrieved from our analysis suggests the possibility of a generic level difference rather than one at the subfamily-level; however, it also

Table 2.4 Procrustes distance in *Atrypa*, *Gotatrypa*, and *Pseudoatrypa* between time units

Gotatrypa	Early Silurian	Middle Silurian	Late Silurian
Early Silurian	0	1.1262	1.7833
Middle Silurian	1.1262	0	1.6767
Late Silurian	1.7833	1.6767	0
Atrypa	Middle Silurian	Late Silurian	Early Devonian
Middle Silurian	0	1.0888	0.70276
Late Silurian	1.0888	0	0.86955
Early Devonian	0.70276	0.86955	0
Pseudoatrypa	Middle Devonian	Late Devonian	
Middle Devonian	0	0.992	
Late Devonian	0.992	0	

confirms that these two subfamilies are closely related. Also, the greater phylogenetic distance between atrypin genera and *Pseudoatrypa* than between atrypin genera and *Spinatrypa* may raise doubts about their phylogenetic arrangement with respect to atrypin genera as retrieved from the phylogenetic chart proposed by Copper (1973). His (Copper 1973) distinctions were made on characters other than shell shape, such as morphology of the pedicle, rib structure and internal features, which may be more diagnostic than simple shell shape. Indeed, the differences in mean shell shape that we found are largely congruent with his (Copper 1973) divisions. In other words, shape analysis has partial bearing on classification, and without including other morphological characters into the morphometric analysis, it is challenging to firmly support the correctness of the phylogeny proposed by Copper (1973). Nevertheless, while the atrypid genera might not have been taxonomically oversplit, the large morphological overlap in shell shape raises questions about the level of distinction among the genera. Overall, though morphometric shell shape is a simple morphological measure, it involves multivariate phenotypic traits that can model complex parts of morphologies in genera, further capturing functional constraints among them, thus proving their usefulness in taxonomic studies (Webster and Zelditch 2009).

2.5.2 Temporal Variation

Phenotypic traits have been studied in the paleontological fossil record, both qualitatively and quantitatively, in terms of long-term and short-term geologic intervals. No one has quantified morphological characters to study atrypids in long term intervals using geometric morphometrics. Our current understanding is that atrypids originated in the late Middle Ordovician (Llandeilo) and increased in generic diversity and abundance during the Early Silurian (upper Llandovery to Wenlockian), after which they declined in diversity during the Late Silurian (Pridoli) through the Early Devonian (Lockhovian), followed by another peak in diversity during the Emsian-Givetian when many of these genera had worldwide distributions

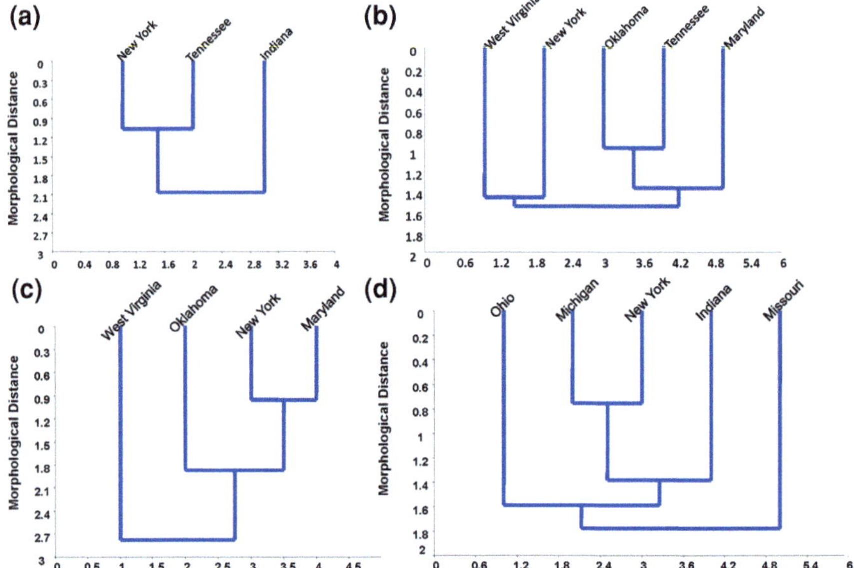

Fig. 2.9 Morphological links for dorsal valves of atrypids in various eastern North America biogeographic locations—**a** *Atrypa* in Middle Silurian; **b** *Atrypa* in Early Devonian; **c** *Kyrtatrypa* in Early Devonian; **d** *Pseudoatrypa* in Middle Devonian

(Copper 2001a, b). Thus, abundant and well preserved atrypids in the Silurian-Devonian geologic interval comprise a great case study to test temporal change.

Our data include atrypid subfamilies and the genera within those subfamilies available for investigation from a 64 Myr (Silurian-Devonian) time period. Using geometric morphometrics, in studying temporal variation patterns, besides solving taxonomic and phylogenetic problems in Silurian-Devonian well preserved, abundant atrypids, is an entirely novel approach. While morphometric differences existed between atrypid subfamilies and genera and for genera between successive time intervals corresponding to the EE subunits ($p < 0.01$; Figs. 2.4, 2.8; Table 2.2), considerable morphological overlap between lowermost and uppermost mean morphological shape occurrences is exhibited with short term changes within lineages in the intermediate time intervals (Fig. 2.8). Overall, smaller morphological distances between atrypids show that the three subfamilies, Atrypinae, Variatrypinae and Spinatrypinae did not differ much based on morphological shape. However, the statistically significant valve shapes differences within these subfamilies over time (Fig. 2.4) could have been in response to their adapting to changing paleoenvironmental conditions prevailing in those time periods. On average, atrypids show smaller average morphological distances in time (0.025–0.028), which is representative of little or no morphological change, as expected in an EEU.

The small magnitude of morphological distance between subfamilies (0.12–1.35), although relatively greater than those between genera within Atrypinae (0.01–0.05), concurs with the current classification system in atrypides.

Table 2.5 Procrustes distance in *Atrypa* from middle Silurian and early Devonian localities, *Kyrtatrypa* from early Devonian localities, and *Pseudoatrypa* from middle Devonian localities within the Eastern North American province

Atrypa (Middle Silurian)	Indiana		New York		Tennessee
Indiana	0		2.2978		1.7388
New York	2.2978		0		1.78
Tennessee	1.7388		1.78		0

Atrypa (Early Devonian)	Maryland	New York	Oklahoma	Tennessee	West Virginia
Maryland	0	1.8611	1.4005	1.3523	2.0355
New York	1.8611	0	1.2749	1.5206	1.8178
Oklahoma	1.4005	1.2749	0	0.99102	1.5379
Tennessee	1.3523	1.5206	0.99102	0	1.2561
West Virginia	2.0355	1.8178	1.5379	1.2561	0

Kyrtatrypa (Early Devonian)	Maryland	New York	Oklahoma	West Virginia
Maryland	0	1.4029	2.0997	2.4742
New York	1.4029	0	1.5751	2.8275
Oklahoma	2.0997	1.5751	0	2.8087
West Virginia	2.4742	2.8275	2.8087	0

Pseudoatrypa (Middle Devonian)	Indiana	Michigan	Missouri	New York	Ohio
Indiana	0	1.463	2.2745	1.6567	1.7556
Michigan	1.463	0	1.4886	0.64741	1.5887
Missouri	2.2745	1.4886	0	1.3663	2.0634
New York	1.6567	0.64741	1.3663	0	1.4282
Ohio	1.7556	1.5887	2.0634	1.4282	0

Surprisingly, the morphological shape distances within *Atrypa*, *Gotatrypa* and *Pseudoatrypa* in time (0.7–1.8) are similar or greater than those measured between subfamilies (Table 2.4). This suggests that this difference may be either due to differences in sample size (as the between-genera distances are based on several samples of the same family, whereas the distances through time are based on sub-samples of the same genus), or it may be due to numerous real short-term changes within the lineages, which get averaged out when comparisons are made between genera. This observed pattern is consistent with stasis. This further suggests that within group variation was greater than between group variation.

2.5.3 Spatial Variation

Morphological distances and range of these distances between geographic localities within *Atrypa* (Middle Silurian and Early Devonian), *Kyrtatrypa* (Early

Devonian) and *Pseudoatrypa* (Middle Devonian) (0.6–2.8) are similar to greater than those observed for individual genus (*Atrypa, Gotatrypa and Pseudoatrypa*) in time (0.7–1.8) (Tables 2.4, 2.5), thus, further confirming greater amount of within-group variation in atrypids. The distances for all these genera also tend to overlap, but the random clustering of geographic localities for each genus and all genera, do not provide a strong biogeographic signal. On average, the smaller average morphological distances in space units (0.027–0.040), suggest little or no morphological change in atrypids spatially.

2.5.4 Environmental Effect

Atrypin, variatrypin and spinatrypin genera lived in broad depositional settings ranging from siliciclastics to carbonates to mixed siliciclastic-carbonate settings (Jodry 1957; Droste and Shaver 1975; Cuffey et al. 1995). For example, *Joviatrypa* preferred quiet, relatively deeper water, muddy substrate assemblage, *Dihelictera* are known to have been derived from a patch reef assemblage, and *Endrea* are derived from biostromal to reefal units (Copper 1995, 1997), while many smooth to tubular ribbed atrypids preferred high energy reefal settings (Copper 1973). *Spinatrypa* have been commonly found in high energy sandy environments (Leighton 2000), though they have also been accounted from low energy muddy environments (Copper 1973). Some of the atrypids of the Genshaw Formation of Traverse Group lived in the full range of rough to quieter energy conditions (McIntosh and Schreiber 1971). Thus, there is a wide variation in preference of substrates and energy conditions for atrypid genera to thrive.

Atrypides expanded in diversity and abundance through Emsian to Givetian, and the expansion of reef growth both equatorially and latitudinally could explain their distribution along the shallow water (<100 m deep) tropical shelf environments (Copper 2001a, b). Overall, atrypin genera preferred nearshore to slope habitats (Zhang and Barnes 2002; Copper 2001a, b), variatrypin genera preferred middle to outer platform habitats (Day 1995) and spinatrypins preferred outer platform to platform margin habitats (Leighton 2000).

While it is probable that the variation observed within individual atrypid genera in time and space suggests some short term changes and within group variation, it is also important to take into account the paleoenvironmental settings from which these genera were derived that may have caused this variation. Testing for preference of habitats, sedimentology, grain size, and lithology and their correlation with respective atrypid genera may provide a clue for the causes behind the morphological variation observed both within a genus and between the genera in time and space. However, this study mainly focuses on testing the taxonomy and phylogeny using geometric morphometrics, and thus the environmental parameters that may have caused this morphological shape change is the scope of a future study.

Results from one test performed were analysed to determine the morphological shape response to lithologic settings. The same genus (*Atrypa, Gotatrypa, Endrea,*

Pseudoatrypa) tested for morphometric shape from shale and carbonate lithologic settings show statistically significant results ($p < 0.01$). However, the temporal patterns observed in genera from three pairs of distinct lithologic settings, Early Silurian packstones and mudstones and Middle Silurian carbonate-silicilastics, Early Silurian packstones and mudstones and Late Silurian carbonate-silicilastics, Middle Devonian carbonate-siliciclastic and Late Devonian carbonates show a slightly higher range of morphological distances (0.99–1.78 Procrustes units) as compared to those from three pairs of similar lithologic settings, Middle and Late Silurian carbonate-siliciclastics, Middle Silurian and Early Devonian carbonate-siliciclastics, and Late Silurian and Early Devonian carbonate-siliciclastics which show a relatively smaller range of morphological distances (0.70–1.09 Procrustes units). Overall, there appears to be no significant relationship between the temporal patterns and lithologic settings, as similar magnitudes (1.1 Procrustes units) of morphological distances result when both overlapping (Middle and Late Silurian carbonate-siliciclastics) and distinct (early Silurian packstones to mudstones and middle Silurian carbonate-siliciclastics) lithologic settings were compared in time. Thus, lithological distribution does not explain for the greater similarity in certain time intervals (middle Silurian-early Devonian: 0.70 Procrustes units, late Silurian-early Devonian: 0.87 Procrustes units) than other intervals (middle Silurian-late Silurian: 1.09 Procrustes units). Similarly, spatial patterns and lithologic settings for genera from the Middle Silurian, Early and Late Devonian time periods exhibit no significant relationship. Magnitudes of morphological distances are similar for both overlapping and distinct lithologies in spatial units. In terms of biogeographic setting, only Early Devonian *Atrypa* genus shows biogeographic signal in that Tennessee and Oklahoma samples are closely linked with more resemblance to Maryland samples than to New York and West Virginia samples. However biogeographically closely spaced Maryland and West Virginia, and Maryland and New York samples from Early Devonian show less similarity. Other genera from Middle Silurian, Early and Middle Devonian (*Atrypa, Kyrtatrypa, Pseudoatrypa*), show no biogeographic signal. For instance, the greater morphological distances between the Middle Devonian closely spaced Missouri and Indiana-Ohio samples as compared to smaller distances between the distantly spaced Missouri and West Virginia-New York samples, suggest that these derived morphological links cannot be explained by biogeographical setting. Neither can these discrepancies be explained by environmental parameters like lithologic settings. Thus, given that the relationship between change in morphological shape and change in other environmental parameters remain unravelled, the morphometric differences in shape observed within genera, between genera, and between subfamilies in time and space could be attributed to their adaptability to other changing environmental conditions or their differential life habits. An overall morphological shape overlap within these groups in the Silurian and Devonian time intervals suggests a close relationship among the genera and subfamilies.

Considerable mean morphological shape overlap between Lower Clinton and Genesee EESU (both relatively close to the P3 EEU boundaries), is indicative of similar climatic settings during this time, such as lowering of the sea level, and the onset of cold climate. However, it is noteworthy that these were different genera in the respective EESUs (*Atrypa, Gotatrypa* represent Lower Clinton and *Pseudoatrypa* represent Genesee) and that they still show considerable overlap, which was probably because they belonged to closely related subfamilies.

If the hypothesis of ecological locking (Morris 1995; Morris et al. 1995) within EEU is correct, then morphological stability is expected within these atrypids as these were sampled from the P3 EEU of the Phanerozoic rock record. Atrypides, most likely, maintained their evolutionary stability through ecological interactions within the unit and as there were no major extinction events within that period, the ecosystem must have remained stable throughout with the exception of minor extinction events that separated the EESUs within the P3 EEU. In this study, overall morphological shape overlap observed between atrypid genera in time, though new taxonomic entities replace each time unit, can be referred to as a case of loose stasis. However, individual genera show large amount of distances between time units and thus, evolutionary rates and modes of each genus in time need to be further investigated to confirm whether loose stasis was really the case.

Morphological evolutionary patterns tested in a few atrypid genera (*Atrypa, Gotatrypa, Pseudoatrypa*) suggest morphological change observed within each lineage is not dramatic as they show some change around the mean which get averaged out in time when compared to other generic pair distances, a pattern similar to stasis. In contrast, these short term changes may be a causal effect of ecophenotypic variation. A few atrypid genera (*Atrypa, Kyrtatrypa, Pseudoatrypa*) tested in space, also show variation within the same group in space units, and no strong biogeographic signal can be derived from the pattern of clustering observed in eastern North American geographic localities. Overall, atrypids are phenotypically plastic, and often distinguishing one genus from another may be very challenging based on morphological shape alone.

2.6 Conclusion

Morphological distances between subfamilies were greater than those between genera within a subfamily, thus suggesting the correct reference of these genera to their respective subfamilies. Evolutionary divergence times among genera retrieved from the phylogenetic tree proposed by Copper (1973) are consistent with the pairwise distances calculated from our morphological shape data, which further supports the taxonomic arrangement and phylogenetic patterns reported in his (Copper 1973) research article. Evolutionary rate and mode indicate that diversifying selection has probably been acting on these atrypid genera at a very slow rate, despite the very small morphological divergences among them. However, some discrepancy arises, and so further evaluation of phylogenetic distances between atrypin genera with that of *Pseudoatrypa* and *Spinatrypa* as a test of relatedness is necessary. Moreover, the morphological shape distances between variatrypin and atrypin genera were so small that these are more like the generic level differences than subfamily level differences. Thus, these discrepancies needs to be further examined through a phylogenetic analysis using the combination of internal morphological features and quantified morphological shape.

Morphological shape analysis shows considerable overlap in Silurian–Devonian atrypid members within the P3 EEU, representing a case of loose stasis. Moreover, large morphological distances between time units within the same genus suggest the possibility of short term changes within a lineage being averaged out when compared with generic pair distances, representing a pattern similar to stasis. Results from several geometric morphometric techniques (including Procrustes analysis and principal component analysis) suggest a certain degree of morphological variability between subfamilies and genera in time and space, which can be attributed to changing paleoenvironmental conditions. Temporal change with some constraints within individual genus (*Atrypa*, *Gotatrypa* and *Pseudoatrypa*) and geographic variation within some genera (*Atrypa*, *Kyrtatrypa* and *Pseudoatrypa*) suggest within group variation was greater than between group variation. Geographic differentiation in morphological shape within atrypids appears to be greater than temporal variation.

Overall, morphological shape change pattern and morphometric divergence in atrypid genera is consistent with the phylogeny proposed by Copper in 1973. Thus, in the 64 myr time scale within the P3 EEU, atrypids in general reflect a high degree of morphological shape conservation in the Silurian-Devonian time interval, regardless of their distinct taxonomic entities.

References

Alvarez F (2006) Forty years since Boucot, Johnson and Staton's seminal paper "On some atrypoid, retzioid, and athyridoid Brachiopoda". Paleoworld 15:135–149

Berry WBN, Boucot AJ (1970) Correlation of the North American Silurian rocks. Geol Soc Am Spec Pap 102:1–289

Bookstein FL (1987) Random walk and the existence of evolutionary rates. Paleobiology 13: 446–464

Bookstein FL (1991) Morphometric tools for landmark data: geometry and biology. Cambridge University Press, Cambridge, p 435

Boucot AJ (1983) Does evolution take place in an ecological vacuum? J Paleontol 57:1–30

Boucot AJ (1986) Ecostratigraphic criteria for evaluating the magnitude, character and duration of bioevent. In: Walliser OH (ed) Global bio-events: lecture notes earth science, vol 8. Springer-Verlag, Berlin, pp 25–45

Boucot AJ (1990a) Phanerozoic extinctions: How similar are they to each other? In: Kauffman EG, Walliser OH (eds) Extinction events in earth history: lecture notes in earth sciences, vol 30. Springer-Verlag, Berlin, pp 5–30

Boucot (1990b) Silurian and pre-upper Devonian bioevents. In: Kauffman EG, Walliser OH (eds) Extinction events in earth history: lecture notes in earth sciences, vol 30. Springer-Verlag, Berlin, pp 125–132

Boucot (1990c) Community evolution: its evolutionary and biostratigraphic significance. In: Miller W (ed) Paleocommunity temporal dynamics: the long-term development of multispecies assemblages, vol 5. Paleontological Society Special Publication, New York, pp 48–70

Boucot AJ, Johnson JG, Staton RD (1964) On Some Atrypoid Retzioid, and Athyridoid Brachiopoda. J Paleontol 38:805–822

Boucot AJ (1975) Evolution and extinction rate controls. Elsevier Scientific Publishing Company, Amsterdam, p 427

Boucot AJ, Blodgett RB (2001) Silurian-Devonian biogeography. In: Brunton HC, Cocks RM, Long SL (eds) Brachiopods Past and present, the natural history museum. Taylor and Francis Publishers, London, pp 335–344

Brett CE, Miller KB, Baird GC (1990) A temporal hierarchy of paleoecological processes within a Middle Devonian epeiric sea. In: Miller W (ed) Paleocommunity temporal dynamics: the long-term development of multispecies assemblages, Paleontological Society Special Publication, New York, pp 178–209

Brett CE, Baird GC (1995) Coordinated stasis and evolutionary ecology of Silurian to middle Devonian faunas in the Appalachian basin. In: Erwin DH, Anstey RL (eds) New approaches to speciation in the fossil record. Columbia University Press, New York, pp 285–315

Brett CE, Ivany LC, Bartholomew AJ, Desantis MK, Baird GC (2009) Devonian ecological-evolutionary subunits in the Appalachian Basin: a revision and a test of persistence and discreteness. Geol Soc Lond 314:7–36 Special Publications

Copper P (1967) Pedicle morphology in Devonian atrypid brachiopods. J Paleontol 41: 1166–1175

Copper P (1973) New Siluro-Devonian atrypoid brachiopods. J Paleontol 47:484–500

Copper P (1977) The Late Silurian brachiopod genus Atrypoidea. Geologiska Foreningers in Stockholm Forhandligas 99:10–26

Copper P (1995) Five new genera of late Ordovician-early Silurian brachiopods from Anticosti island Eastern Canada. J Paleontol 69:846–862

Copper P (1997) New and revised genera of Wenlock-Ludlow Atrypids (Silurian Brachiopoda) from Gotland Sweden, and the United Kingdom. J Paleontol 70:913–923

Copper P (1996) New and revised genera of Wenlock-Ludlow Atrypids (Silurian Brachiopoda) from Gotland Sweden, and the United Kingdom. J Paleontol 70:913–923

Copper P (2001a) Radiations and extinctions of atrypide brachiopods: ordovician–Devonian. In: Brunton CHC, Cocks LRM, Long SL (eds) Brachiopods past and present. Natural History Museum, London, pp 201–211

Copper P (2001b) Reefs during the multiple crises towards the Ordovician-Silurian boundary Anticosti Island, eastern Canada, and worldwide. Can J Earth Sci 38:153–171

Copper P (2002) Atrypida. In: Kaesler RL (ed) Brachiopoda (revised), part H of treatise on invertebrate paleontology. The Geological Society of America Inc. and the University of Kansas, Boulder, Colorado and Lawrence, Kansas, pp 1377–1474

Copper P (2004) Silurian (Late Llandovery-Ludlow) Atrypid Brachiopods from Gotland, Sweden, and the Welsh Borderlands, the Great Britain. National Research Council of Canada Research Press, Ottawa

Cuffey CA, Robb AJIII, Lembcke JT, Cuffey RJ (1995) Epizoic bryozoans and corals as indicators of life and post-mortem orientations of the Devonian brachiopod *Meristella*. Lethaia 28:139–153

Cumings ER, Shrock RR (1928) The Silurian coral reefs of northern Indiana and their associated strata. Proc Indian Acad Sci 36:71–85

Day J (1995) Brachiopod fauna of the Upper Devonian (late Frasnian) Lime Creek Formation of north-central Iowa, and related units in eastern Iowa. In: Bunker BJ (ed) Geological society of Iowa guidebook, vol 62. Iowa City, 21–40

Day J (1998) Distribution of latest Givetian-Frasnian Atrypida (Brachiopoda) in central and western North America. Acta Palaeontol Pol 43:205–240

Day J, Copper P (1998) Revision of latest Givetian-Frasnian Atrypida (Brachiopoda) from central North America. Acta Palaeontol Pol 43:155–204

Droste JB, Shaver RH (1975) Jeffersonville limestone (middle Devonian) of Indiana Stratigraphy, sedimentation, and relation to Silurian reef-bearing rocks. Am Assoc Petrol Geol Bull 59:393–412

Felsenstein J (1988) Phylogenies and quantitative characters. Annu Rev Ecol Syst 19:445–471

Fenton CL, Fenton MA (1930) Studies on the genus *Atrypa*. Am Midl Nat 12:1–18

Gingerich PD (1993) Quantification and comparison of evolutionary rates. Am J Sci 293A: 453–478

Holterhoff PE (1996) Crinoid biofacies in upper Carboniferous cyclothems, midcontinent North America: faunal tracking and the role of regional processes in biofacies recurrence. Palaeogeogr Palaeoclimatol Palaeoecol 127:47–81

Hunt G (2007) The relative importance of directional change, random walks, and stasis in the evolution of fossil lineages. Proc Nat Acad Sci U S A 104:18404–18408

Ivany LC, brett CE, Wall HLB, Wall PD, Handley JC (2009) Relative taxonomic and ecologic stability in Devonian marine faunas of New York State: a test of coordinated stasis. Paleobiology 35:499–524

Jodry RL (1957) Reflection of possible deep structures by traverse group facies changes in western Michigan. Am Assoc Petrol Geol Bull 41:2677–2694

Koch W, Day J (1995) Late Eifelian-early Givetian (Middle Devonian) brachiopod paleobiogeography of eastern and central North America: brachiopods. In: Proceedings of the third international brachiopod congress, vol 3. pp 135–143

Lande R (1976) Natural selection and random genetic drift in phenotypic evolution. Evolution 30:314–334

Leighton LR (2000) Environmental distribution of spinose brachiopods from the Devonian of New York test of the soft-substrate hypothesis. Palaios 15:184–193

Lieberman BS, Brett CE, Eldredge N (1995) A study of stasis and change in two species lineages from the middle Devonian of New York state. Paleobiology 21:15–27

Lowenstam HA (1957) Niagaran reefs in the Great Lakes area. Geol Soc Am Memoir 67:215–248

Macleod N (2002) Geometric morphometrics and geological shape-classification systems. Earth Sci Rev 59:27–47

Mcintosh GC, Schreiber RL (1971) Morphology and taxonomy of the middle Devonian crinoid *Ancyrocrinus bulbosus* Hall 1862: contributions from the museum of paleontology, vol. 23. The University of Michigan, Ann Arbor, pp 381–403

Morris PJ (1995) Coordinated stasis and ecological locking. Palaios 10:101–102

Morris PJ, Ivany LC, Schopf KM, Brett CE (1995) The challenge of paleoecological stasis: Reassessing sources of evolutionary stability. Proc Natl Acad Sci 92:11269–11273

Polly PD (2003) Paleophylogeography: the tempo of geographic differentiation in marmots (Marmota). J Mammal 84:369–384

Polly PD (2004) On the simulation of morphological shape: mutivariate shape under selection and drift. Palaeontologia Electronica 7(7A):1–28

Polly PD (2008) Adaptive zones and the Pinniped ankle: a three-dimensional quantitative analysis of carnivoran tarsal evolution. In: Sargis E, Dagosto M (eds) Mammalian evolutionary morphology: a tribute to Frederick S. Szalay. Springer, Dordrecht, pp 167–196

Rode AL, Lieberman BS (2005) Integrating evolution and biogeography: a case study involving Devonian crustaceans. J Paleontol 79:267–276

Rohlf FJ (1990) Rotational fit (Procrustes) methods. In: Rohlf FJ, Bookstein FL (eds) Proceedings of the Michigan morphometrics workshop. pp 227–236

Rohlf FJ (1999) Shape statistics Procrustes superimpositions and tangent spaces. J Classif 16:197–223

Rohlf FJ, Slice DE (1990) Extensions of the Procrustes method for the optimal superimposition of landmarks. Syst Zool 39:40–59

Rohlf FJ, Marcus LF (1993) A revolution in morphometrics. Trends Ecol Evol 8:129–132

Rohlf FJ (2004) tpsDig; Version 1.4. Department of Ecology and Evolution, State University of New York at Stony Brook.tpsSplin. URL:http://life.bio.sunysb.edu/morph/. Accessed 15 Aug 2004

Roopnarine PD (2001) The description and classification of evolutionary mode: a computational approach. Palaeobiology 27:446–465

Roopnarine PD (2003) Analysis of rates of morphologic evolution. Annu Rev Ecol Evol Syst 34:605–632

Slice DE (2001) Landmark coordinates aligned by Procrustes analysis do not lie in Kendall's shape space. Syst Biol 50:141–149

Sheehan PM (1991) Patterns of synecology during the Phanerozoic. In: Dudley EC (ed) The unity of evolutionary biology. Dioscorides Press, Portland, pp 103–118

Sheehan PM (1996) A new look at ecologic evolutionary units (EEUs). Palaeogeogr Palaeoclimatol Palaeoecol 127:21–32

Van der Pluijm B (1993) Paleogeography, accretionary history, and tectonic scenario: a working hypothesis for the Ordovician and Silurian evolution of the Northern Appalachians. Geol Soc Am Spec Pap 275:27–40

Webster M, Zelditch ML (2009) Testing hypotheses of developmental constraints on macroevolutionary diversification: studying modularity of ancient developmental systems, vol 3. Cincinnati Museum Center Scientific Contributions, Cincinnati, p 115

Webster M (2011) The structure of cranidial shape variation in three early ptychoparioid trilobite species from the Dyeran–Delamaran (traditional "lower–middle" Cambrian) boundary interval of Nevada, U. S. A. J Paleontol 85:179–225

Williams A, Rowell AJ (1965) Brachiopoda (revised), part H, vol. 2 of Treatise on Invertebrate Paleontology. In: Moore RC (ed) Geological Society of America and Lawrence, Boulder, The University of Kansas, Lawrence, p 927

Williams A, Brunton CHC, Carlson SJ (2002) Brachiopoda (revised), part H, vol. 4 of Treatise on Invertebrate Paleontology. In: Kaesler RL (ed) Geological Society of America and Lawrence, Boulder, The University of Kansas, Lawrence, p 807

Zelditch ML, Swiderski DL, Sheets HD, Fink WL (2004) Geometric morphometrics for biologists, Elsevier, Academic Press, New York

Zhang S, Barnes CR (2002) Paleoecology of Llandovery conodonts Anticosti Island, Quebec. Palaeogeo Palaeoclimatol Palaeoecol 180:33–55

Chapter 3
Morphological Evolution in an Atrypid Brachiopod Lineage from the Middle Devonian Traverse Group of Michigan, USA: A Geometric Morphometric Approach

3.1 Introduction

The mode and rate of morphological change in lineages over geologic time has been a hotly debated topic in paleontology and biology over the last three decades. The punctuated equilibrium model (Eldredge and Gould 1972; Gould and Eldredge 1977; Stanley 1979) has had a major impact on the study of Paleozoic faunas, and many studies have confirmed punctuated patterns of change in Paleozoic species. Lieberman et al. (1995) found a pattern of punctuated stasis in brachiopods from the Middle Devonian Hamilton Group of the Appalachian Basin. In this study, we examine contemporary brachiopods from the Traverse Group of the Michigan Basin to see whether the patterns found in the Appalachian Basin are regional or local in scope.

The paleontological record of the lower and middle Paleozoic Appalachian foreland basin demonstrates ecological and morphological stability on geological time scales (Brett and Baird 1995). Some 70–80 % of fossil morphospecies within assemblages persisted in similar relative abundances in coordinated packages lasting as long as seven million years despite evidence for environmental change and biotic disturbances (Morris et al. 1995). This phenomenal evolutionary stability despite environmental fluctuations has been explained by the concept of ecological locking. Ecological locking provides the source of evolutionary stability that is suggested to have been caused by ecological interactions that maintain a static adaptive landscape and prevent both the long-term establishment of exotic invading species and evolutionary change of native species (Morris 1995; Morris et al. 1995). For example, competition plays an important role in mediating stasis by stabilizing selection (Lieberman and Dudgeon 1996). Though studying the community stability patterns is beyond the scope of this study, it would be interesting to examine whether similar morphological stability is exhibited in the Michigan Basin Traverse group setting.

R. Bose, *Biodiversity and Evolutionary Ecology of Extinct Organisms,* Springer Theses, DOI: 10.1007/978-3-642-31721-7_3, © Springer-Verlag Berlin Heidelberg 2013

Models of stasis and gradual change have previously been tested in Devonian taxa from the Appalachian Basin. Isaacson and Perry (1977) did not find any significant change in *Tropidoleptus carinatus*, an orthide brachiopod from the Givetian of the Hamilton Group from the lowest to its highest occurrence, spanning some 40 m.y. Goldman and Mitchell (1990) tested the internal morphology of three brachiopod species of the Hamilton Group of western New York from size measurements and found only one species of Late Givetian age showed some species level change. Eldredge further tested the same fauna studied by Goldman and Mitchell (1990) using morphometrics and found almost no significant morphological change in this unit (Brett and Baird 1995). Later workers (Lieberman et al. 1995) tested two common brachiopod species lineages using size measurements on the pedicle valves of 401 *Mediospirifer audaculus* and 614 *Athyris spiriferoides* from successive stratigraphic horizons in the Hamilton Group, a section which has a five million year duration. They found morphological overlap within these species between the lowermost and uppermost strata with some variations in the intervening samples of the Hamilton Group and concluded that the pattern of change was one of stasis. All of the Hamilton brachiopod species lineages that were studied showed stasis or, at most, minor evolutionary changes.

This study tests the hypothesis of stasis in the Michigan Basin, a biogeographically separate setting from the Appalachian Basin. These two basins are separate sub-provinces of Eastern North America, formerly a part of the eastern Laurentian paleocontinent. Michigan Basin is a large intracratonic basin of Eastern North America situated south of the Canadian Shield, containing Paleozoic sedimentary rocks (Cambrian-Carboniferous) and a thin surface layer of Jurassic sediments at the center of the basin. The basin is centered in Michigan's Southern Peninsula including parts of Michigan's Northern Peninsula, northern and eastern edges of Wisconsin, northeastern Illinois, northern parts of Indiana and Ohio, and extreme western part of Ontario, Canada. The Appalachian Basin is a foreland basin situated southeast of the Michigan Basin, containing Paleozoic sedimentary rocks and extending from New York, Pennysylvania, eastern Ohio, West Virginia, western Maryland in the north to eastern Kentucky, western Virginia, eastern Tennessee, northwestern Georgia, and northeastern Alabama in the south. The Findlay–Algonquin Arch separates the intracratonic basin from the foreland basin, which were connected by shallow seas to varying degrees during the late Eifelian to Givetian interval (Bartholomew and Brett 2007; Brett et al. 2010).

Based on biostratigraphic evidence and sequence stratigraphic analysis, the Traverse EE subunit correlates with the Hamilton EE subunit (Brett and Baird 1995; Bartholomew 2006; Brett et al. 2009, 2010). The Traverse fauna within the Middle Devonian of the Michigan Basin subprovince displays a high level of faunal and compositional persistence and thus is defined as the Traverse EE subunit (Bartholomew 2006). The small-scale community stability of the Hamilton fauna within the Middle Devonian of the Appalachian Basin subprovince is defined as the Hamilton EE subunit (Brett and Baird 1995). Thus, the morphological trends observed in this study of the Traverse Group can be compared to previously studied patterns in the contemporary Hamilton Group.

(a)   **(b)**

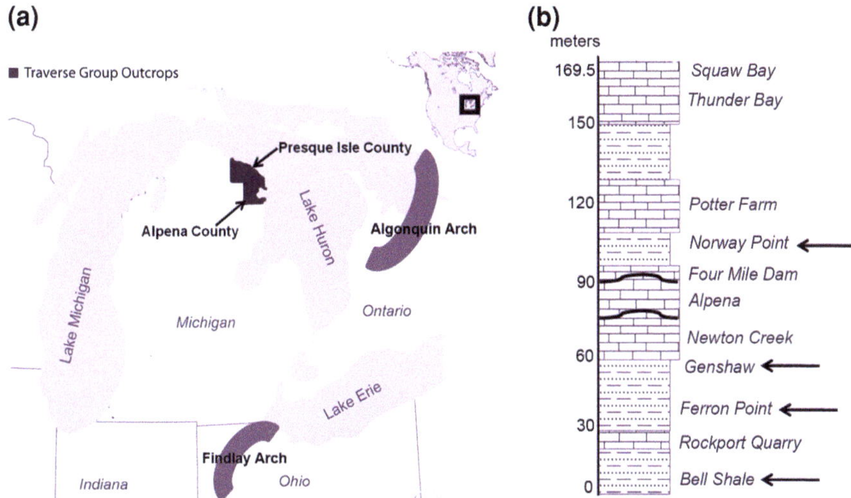

Fig. 3.1 **a** Map showing location of Michigan quarries of the Middle Devonian Traverse Group, namely the Alpena and Presque Isle Quarries, from which the samples used in this study were collected, **b** simplified stratigraphic section of the Traverse Group at Alpena Quarry showing the eleven stratigraphic intervals exposed on the northeastern outcrop of Michigan. *Arrows* show the four stratigraphic intervals from where the specimens used in this study were collected. *Curved lines* show the location of unconformities (after Wylie and Huntoon 2003)

In this study, changes in morphological shape over time were assessed in the *Pseudoatrypa cf. lineata* lineage from a 5 m.y. long section of the Givetian in the Traverse Group. The brachiopod species *Pseudoatrypa cf. lineata* (Webster 1921) was subjected to geometric morphometric and multivariate statistical analyses to examine mode and rate of morphological shape evolution. This species was sampled from four richly fossiliferous formations in the Alpena and Presque Isle Counties of the northeastern outcrop of Michigan: Bell Shale, Ferron Point, Genshaw and Norway Point (Fig. 3.1). The stratigraphy and paleoenvironment of the Traverse Group have been the subject of many detailed studies (e.g., Ehlers and Kesling 1970; Kesling et al. 1974; Wylie and Huntoon 2003), allowing fossil specimens collected from the Traverse Group to be placed in a paleoenvironment setting. These data were used to evaluate the environmental context of these four samples, especially to determine whether changes in water depth are correlated with observed changes in the brachiopods (Wylie and Huntoon 2003).

3.1.1 Hypotheses

(1) If the species lineage *P. cf. lineata* evolved according to the punctuated equilibrium model, in which morphological change occurs predominantly at speciation and otherwise remains static through the rest of its history, then we would expect no significant differences between samples of the species from

successive stratigraphic units of the Middle Devonian Traverse Group over time; (2) if the species evolved in a gradual, directional manner, then we would expect samples close together in time to be more similar to one another than those more separated in time; and (3) if morphological shape was affected by change in environmental factors like water depth, etc. then we would expect a strong correlation between changes in such factors and changes in shell shape.

3.2 Geologic Setting

The geologic setting used to test the proposed hypotheses in the *Pseudoatrypa cf. lineata* species lineage is the Traverse Group, a package of rocks from Michigan that spans roughly 6.0 m.y. of the Middle Devonian (Givetian) and the lower Upper Devonian (Frasnian). The appearance of rocks in North America seems to have been driven by the post-Eifelian augmentation of the Acadian orogeny (Wylie and Huntoon 2003). The Traverse Group and its fauna are associated with the influx of siliciclastic sedimentation from this orogeny (Brett 1986; Cooper et al. 1942; Ettensohn 1985; Ehlers and Kesling 1970; Wylie and Huntoon 2003). Depending upon the frequency of these storm events and the turbidity of the water column from influx of siliciclastic sedimentation from the Taghanic onlap, the faunas in these settings can be influenced by a variety of environmental parameters like light intensity variations, sedimentation rate, dissolved oxygen concentration, salinity, and temperature variations.

The Traverse strata comprise a nearly 169.5 m thick succession of sedimentary rocks, primarily shales, claystones and limestones, which were deposited in predominantly supratidal to nearshore marine settings (Ehlers and Kesling 1970; Wylie and Huntoon 2003). The Bell Shale, about 21.0 m in thickness, consists of a basal crinoid rich lag and shales, which were deposited with water depth ranging from 25.0–44.0 m. The Ferron Point, about 13.0 m in thickness, consists of soft shales and limestones, deposited with water depth approximately 39.0 m. The Genshaw Formation, 35.0 m in thickness, consists of soft shales and argillaceous limestones, with water depth ranging from 25.0 to 39.0 m. Norway Point Formation, 13.5 m in thickness, consists of abundant shales and claystones, with limestones, deposited at approximately, <7.0 m water depth (Wylie and Huntoon 2003). The formations chosen for data collection in this study are dominated by shales (Wylie and Huntoon 2003) and thus for this study, sampling restricted to shale beds in the four formations allows morphological analysis in a more or less stable environmental setting. Regarding water depths, these samples represent nearshore low energy environments from shallow to medium water depths (7.0–50.0 m) which may have been interrupted by occasional storms.

The patterns of shape change in these Michigan Basin samples will be compared with the patterns found by previous workers in the Appalachian Basin from the contemporary Hamilton Group units, which the Traverse units have been correlated based on sequence stratigraphic analysis (Brett et al. 2010). The

Michigan Basin is separated from the Appalachian Basin by a basement arch system, the northeastern segment of which is called the Algonquin Arch, and the southwestern segment called the Findlay Arch (Carlson 1991). The Traverse Group lies on the flanks of the Michigan Basin and my study area is located in the northeastern part (Northern Peninsula) of the Michigan Basin in Alpena and Presque Isle Counties, Michigan. The Hamilton Group lies on the flanks of the Appalachian Basin, with samples studied in the past for stasis coming from western and central New York, northwestern part of Appalachian Basin.

Conodont-based correlation and sequence stratigraphic analysis of the Middle Devonian strata from the Michigan and Appalachian Basin shows that the Bell Shale is coeval with the upper Marcellus Shale, the Ferron Point and Genshaw formations are coeval with the Skaneateles, and the Norway Point Formation with the lower Windom Member of the Moscow Formation (Brett et al. 2009, 2010). Thus the Traverse Group and Hamilton Group fauna thrived during similar geologic time periods within the Hamilton EESU (Orr 1971; Cooper and Dutro 1982; Bartholomew and Brett 2007; Brett et al. 2010).

3.3 Materials and Methods

Bulk atrypid samples were first qualitatively examined and identified to species level based on external morphological characteristics. Thousand hundread and twenty four specimens of *P. cf. lineata* were used from a total of four different shale beds at six localities of four above mentioned strata (Bell Shale = 131, Ferron Point = 330; Genshaw = 506; Norway Point = 157) in Michigan. Most of the samples used in this study are from the Michigan Museum of Paleontology Collections, the rest are from the collections of Alex Bartholomew from State University of New York that have now been deposited at the Indiana University Paleontology Collections.

Seven landmark points were chosen for morphological analysis (Fig. 3.2, Table 3.1), each landmark point representing the same location on each specimen to capture the biologically most meaningful shape. These landmark points are geometrically homologous (sensu Bookstein 1991), which is appropriate for analyses attempting to capture morphological shape changes (Rohlf and Marcus 1993). Brachiopods are bilaterally symmetrical organisms and each side is a mirror image of the other, i.e., each half captures the shape of the organism. Thus, for all individuals, measurements were taken on half of the specimen (dorsal view right, ventral view left, one side of anterior and posterior views) (Fig. 3.2). Data were captured using the TPSdig program for digitizing landmarks for geometric morphometrics. Procrustes analysis (Rohlf 1990, 1999; Rohlf and Slice 1990; Slice 2001) was performed on original shape data, rotating, translating and scaling all landmarks to remove all size effects, while maintaining their geometric relationships (Procrustes superimposition). Principal component analysis was performed to determine the morphological variation between samples from the four

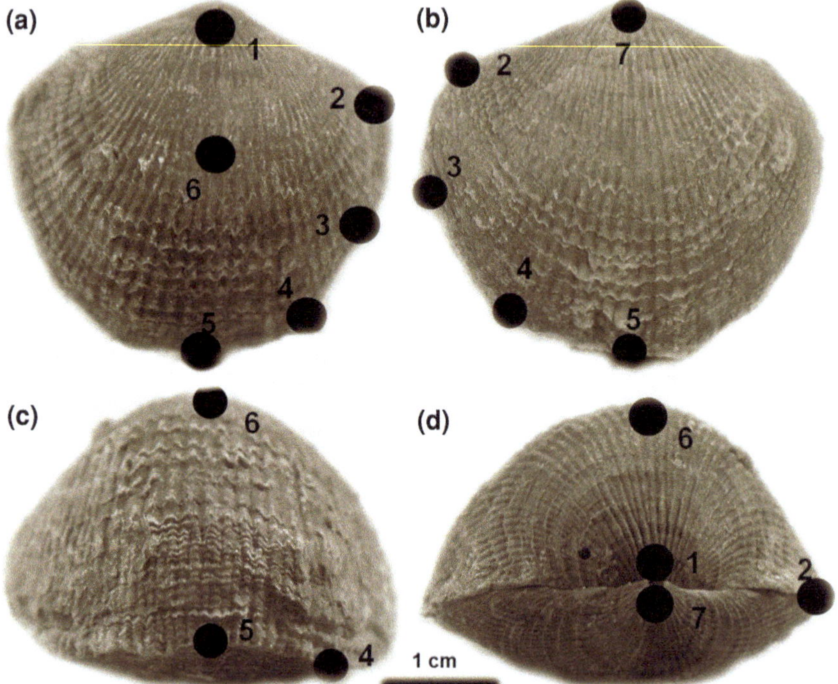

Fig. 3.2 Seven landmark points on *Pseudoatrypa cf. lineata* shells with distribution of **a** six landmark points on *right side* of dorsal valve, **b** five landmark points on *left side* of ventral valve, **c** three landmark points on *right side* of anterior margin area of shell, and **d** four landmark points on *right side* of posterior hinge view of shell

Table 3.1 Landmark points on the *Pseudoatrypa cf. lineata* shell representing geometric positions that are biologically functional

Landmark points	Area of the shell	Landmark descriptions
1	D	Tip of umbo
2	D/V	Junction on the hinge of dorsal valve interarea, ventral valve interarea and commissure
3	D/V	Midpoint of specimen length projected onto commissure, length midpoint based on length of baseline
4	D/V	Extreme edge of anterior commissure adjacent to L5
5	D/V	Edge of commissure perpendicular to hinge, in line with L1 (on sulcate specimens, this point coincided with the lowest point of the sulcus)
6	D	Maximum height of curvature
7	V	Lowest point of interarea/on the pedicle foramen

D dorsal and *V* ventral

Table 3.2 Eleven stratigraphic units of the traverse group showing their thickness in meters and their estimated duration in million years

	Stratigraphic units	Thickness (m)	Time (m.y.)
Traverse group	Squaw bay	2.85	0.10
	Thunder bay	4.8	0.17
	Potter farm	30.6	1.08
	Norway	**13.5**	**0.48**
	Four mile dam	6.3	0.22
	Alpena	23.7	0.84
	Newton creek	7.5	0.27
	Genshaw	**34.95**	**1.24**
	Ferron	**12.6**	**0.45**
	Rockport	12.6	0.45
	Bell	**20.4**	**0.72**
	Total thickness	169.5	6.02

Time in million years for each unit was estimated by the proportional thickness relative to the entire traverse group, which is estimated to have been deposited over 6 million years. *Pseudoatrypa cf. lineata* samples are from the stratigraphic units in bold

stratigraphic intervals using the principal component axes 1 and 2. Multivariate analysis of variance (MANOVA) was performed to test for differences in mean shape between stratigraphic units.

Cluster analysis based on Euclidean distances was performed to illustrate shape similarities among the stratigraphic units. The relationship between change in mean shape and time was used to estimate the rate of change and evolutionary mode. Appropriate absolute age dates are not available for the individual stratigraphic units, so their durations and the intervals of time between them were estimated based on the total thickness and duration of the Traverse Group using the section thicknesses reported by Wylie and Huntoon (2003) (Tables 3.2 and 3.3).

Evolutionary rate and mode in morphological divergence was assessed using maximum-likelihood (Polly 2004, 2008). This method estimates the mean per-step evolutionary rate and the degree of stabilizing or diversifying selection from the time-distance matrix of pairwise morphological distances and divergence times. Morphological distance was calculated as pairwise Procrustes distances among *P. cf. lineata* taxon and divergence time was taken from the estimates of the total time in millions of years between strata that the species have been diverging independently. The method uses the following equation to estimate rate and mode simultaneously,

$$D = rt^{\wedge}a. \tag{3.1}$$

where D is morphological divergence (Procrustes distance), r is the mean rate of morphological divergence, t is divergence time, and a is a coefficient that ranges from 0.0 to 1.0, where 0.0 represents complete stabilizing selection (stasis), 0.5 represents perfect random divergence (Brownian motion) and 1.0 represents perfect diversifying (directional) selection (Polly 2008). Maximum-likelihood is

Table 3.3 Matrix showing time differences (million years) between the four stratigraphic units used for this study estimated from thicknesses of the strata

	Bell	Ferron	Genshaw	Norway
Bell	0	1.16	1.61	4.18
Ferron	1.16	0	0.45	3.02
Genshaw	1.61	0.45	0	2.57
Norway	4.18	3.02	2.57	0

used to find the parameters r and a that maximize the likelihood of the data, and are thus the best estimates for rate and mode. The data were bootstrapped 1,000 times to generate standard errors for these estimates. This method is derived directly from the work presented by Polly (2004, 2008) and is mathematically related to other methods in evolutionary genetics (Felsenstein 1988; Gingerich 1993; Hunt 2007; Lande 1976; Roopnarine 2003).

Morphological shape in *Pseudoatrypa cf. lineata* individuals were tested against the shallow (Bell Shale = 34.35 m, Ferron Point = 39.3 m, Genshaw Fm = 31.95 m) and medium (Norway Point = 1.9 m) classes of water depth with respect to the water depth data from Wylie and Huntoon (2003). Multivariate analysis of variance (MANOVA) was performed between water depths and principal component scores of the shape co-ordinates of the valve morphology of individual shells from the Traverse Group strata to test for a statistical significant relationship between the two. In addition, the mean morphological PC scores for the samples were statistically correlated with water depth using linear regression analysis.

3.4 Results

The material from the Bell Shale, Ferron Point, Genshaw and Norway Point strata of the Traverse Group studied here was referred to *Pseudoatrypa cf. lineata* based on distinctive qualitative characters. Webster (1921) first described this taxon as *Atrypa lineata* from the late Givetian-early Frasnian Cedar Valley Group of Iowa, material which was later recognised as *Pseudoatrypa lineata* species of the new genus *Pseudoatrypa* (Copper 1973; Day and Copper 1998). The holotype of *Atrypa lineata* (Webster 1921) came from the upper Osage Springs and Idlewild members of the Lithograph City Formation of Iowa (Day 1992, 1996). *Pseudoatrypa* also occurs in the Traverse Group of the Michigan Basin (Stumm 1951; Copper 1973; Koch 1978) and in the Silica Formation and equivalent rocks from northern Indiana (Wiedman 1985). Copper (1973) described *Pseudoatrypa* as medium-sized with dorsiconvex to convexiplanar, subtriangular to subrectangular shaped shells. These shells were usually strongly uniplicate with subtubular to sublamellar rib structure, 2–3 mm spaced growth lamellae and either lacking frills or containing very short projecting frills. They have a small ventral beak with a foramen commonly expanding or enlarging into the umbo, with small interarea and tiny deltidial plates. Their internal features contain small dental cavities with delicate tooth and socket plates, disjunct jugal processes and spiralia with 8–12 whorls. *P. lineata* is described as medium- to large-sized (up to 37 mm in length, 35 mm in width) with globose dorsibiconvex-convexiplanar shells and an inflated hemispherical dome-like dorsal valve, shell length exceeding width slightly in all growth stages, subquadrate shell outline, broad to angular fold developed posterior of mid-valve, becoming more pronounced towards anterior margin in large adult shells (30 mm), exterior of both valves with fine radial tubular ribs (9–10/5 mm at anterior margin), regularly spaced concentric lamellae crowding towards anterior and lateral margins in larger adults (20 mm length), short frills rarely preserved (Fenton and Fenton 1935; Day and Copper 1998). *P. lineata* ranged from late Givetian to early Frasnian of North America (Day and Copper 1998).

Samples in this study from all the four stratigraphic intervals of the Givetian age Traverse Group agree well with the overall morphology of *P. lineata* and were referred to *Pseudoatrypa cf. lineata* for the purpose of this study. These samples are characterised by shell maximum width of 2.1−3.3 ± 0.2 cm, width almost equal to the length of the shell, subquadrate shell shape, dorsibiconvex-convexi-planar shell, flattened with/without umbonal inflation in ventral valves, fine to coarse ribs with implantations and bifurcations, 1–2 plicae/1 mm spacing, some-what consistent 2–4 mm spacing between growth lines with their crowding at the anterior margin. However, based on morphometric results derived from quantifi-cation of morphological shape, the Norway Point samples appear to be different from the lower Traverse Group samples.

Principal component axis 1 (PC1) explained 36.0 % of shape variation in dorsal valves, 28.8 % of variation in ventral valves, 62.6 % variation in anterior margin area, and 82.9 % variation in posterior hinge area. PC2 explained 21.6 % of variation in dorsal valves, 23.9 % variation in ventral valves, 35.8 % variation in anterior margin area, and 13.1 % variation in posterior hinge area. There is some shape variation between stratigraphic intervals, mostly in the deviation of the Norway Point sample from the underlying stratigraphic horizons (Fig. 3.3).

MANOVA indicated that differences in mean shape between stratigraphic intervals were significant for dorsal valves (F = 39.58, df1 = 24, df2 = 3336, $p < 0.01$), ventral valves (F = 29.65, df1 = 18, df2 = 3345, $p < 0.01$), anterior (F = 38.72, df1 = 6, df2 = 2228, $p < 0.01$) and posterior (F = 90.48, df1 = 12, df2 = 3339, $p < 0.01$) (Table 3.4). The statistical significance of the MANOVA (post hoc pairwise tests with Bonferroni correction) demonstrates that there is some real differentiation between the samples from the four stratigraphic horizons in shell shape (Table 3.4), but the substantial overlap in variation and the difficulty in visually distinguishing the differences in shell shape suggests to us that all these samples of *P. cf. lineata* species show little morphological change over time (Figs. 3.3 and 3.4). However, abrupt deviation in mean morphological shape of the Norway Point samples (Fig. 3.3) gives some evidence of morphological change in this species later in time.

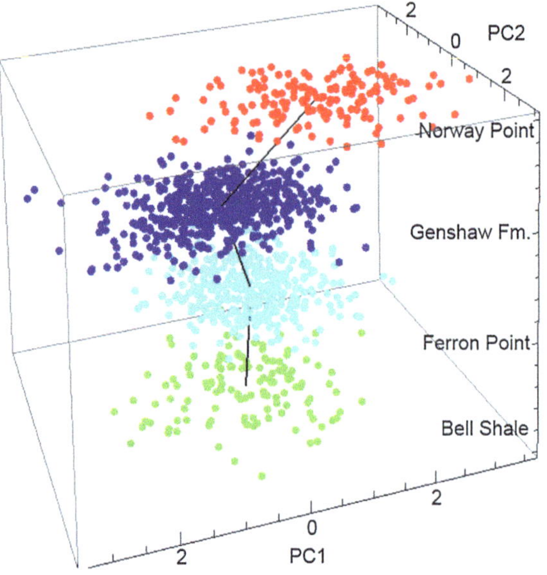

Fig. 3.3 Morphological variation (PC1 = 36.02 % and PC2 = 21.62 %) and mean morphological shape trend in dorsal valves of *P. cf. lineata* along four Traverse Group formations (Bell Shale, Ferron Point, Genshaw Formation and Norway Point)

Morphological differences between stratigraphic samples ranged from 0.47 to 1.97 Procrustes units (the units of difference in the principal components space) (Table 3.5). Closely spaced samples had smaller morphological distances than did widely separated samples. Overall, the morphological distances concur with the stratigraphic succession (Fig. 3.5).

Procrustes pairwise distances were plotted against time calculated from stratigraphic thicknesses and maximum likelihood used to estimate the rate and mode of evolution (Fig. 3.6). Perfectly random Brownian-motion change corresponds to a mode coefficient of 0.5, perfectly directional change to a coefficient of 1.0 and perfect stasis to a coefficient of 0.0. All four data sets had coefficients less than 0.5, indicating that morphological divergence was less than expected with random changes and suggesting that morphology is constrained in a stasis-like pattern. The anterior and posterior landmarks were nearest to stasis, and the ventral valve showed less constraint than the dorsal valve.

MANOVA and regression analysis shows significant difference between medium and shallow water depth brachiopod samples for dorsal valves ($p \ll 0.01$), ventral valves ($p \ll 0.01$), anterior ($p \ll 0.01$) and posterior ($p \ll 0.01$). Water depth correlated only weakly with the individual sample mean principal component scores of *P. cf. lineata* for dorsal valves ($r = -0.14$), ventral valves ($r = 0.33$), anterior (-0.11) and posterior (-0.07).

Table 3.4 '*p*' values show distinctness between *Pseudoatrypa cf. lineata* samples from four stratigraphic units in time for (a) dorsal valve, (b) ventral valve, (c) anterior marginal area, and (d) posterior hinge area ($p < 0.01$)

Dorsal	Bell	Ferron	Genshaw	Norway
Bell	0	0.000	0.000	0.000
Ferron	0.000	0	0.000	0.000
Genshaw	0.000	0.000	0	0.000
Norway	0.000	0.000	0.000	0
Ventral	Bell	Ferron	Genshaw	Norway
Bell	0	0.000	0.000	0.000
Ferron	0.000	0	0.000	0.000
Genshaw	0.000	0.000	0	0.000
Norway	0.000	0.000	0.000	0
Anterior	Bell	Ferron	Genshaw	Norway
Bell	0	0.000	0.000	0.000
Ferron	0.000	0	0.000	0.001
Genshaw	0.000	0.000	0	0.000
Norway	0.001	0.001	0.000	0
Posterior	Bell	Ferron	Genshaw	Norway
Bell	0	0.000	0.000	0.000
Ferron	0.000	0	0.000	0.000
Genshaw	0.000	0.000	0	0.000
Norway	0.000	0.000	0.000	0

Table 3.5 Procrustes pairwise distances for *Pseudoatrypa cf. lineata* lineage between stratigraphic units in time for (a) dorsal valve, (b) ventral valve, (c) anterior marginal area, and (d) posterior hinge area

Dorsal	Bell	Ferron	Genshaw	Norway
Bell	0	0.94315	1.1128	1.9706
Ferron	0.94315	0	1.3173	1.8075
Genshaw	1.1128	1.3173	0	1.8232
Norway	1.9706	1.8075	1.8232	0
Ventral	Bell	Ferron	Genshaw	Norway
Bell	0	0.84557	1.1987	1.686
Ferron	0.84557	0	0.79457	1.7012
Genshaw	1.1987	0.79457	0	1.295
Norway	1.686	1.7012	1.295	0
Anterior	Bell	Ferron	Genshaw	Norway
Bell	0	0.63057	0.75156	0.5672
Ferron	0.63057	0	0.85516	0.47276
Genshaw	0.75156	0.85516	0	1.1393
Norway	0.5672	0.47276	1.1393	0
Posterior	Bell	Ferron	Genshaw	Norway
Bell	0	0.74795	1.1225	1.7207
Ferron	0.74795	0	1.5494	1.8043
Genshaw	1.1225	1.5494	0	1.6468
Norway	1.7207	1.8043	1.6468	0

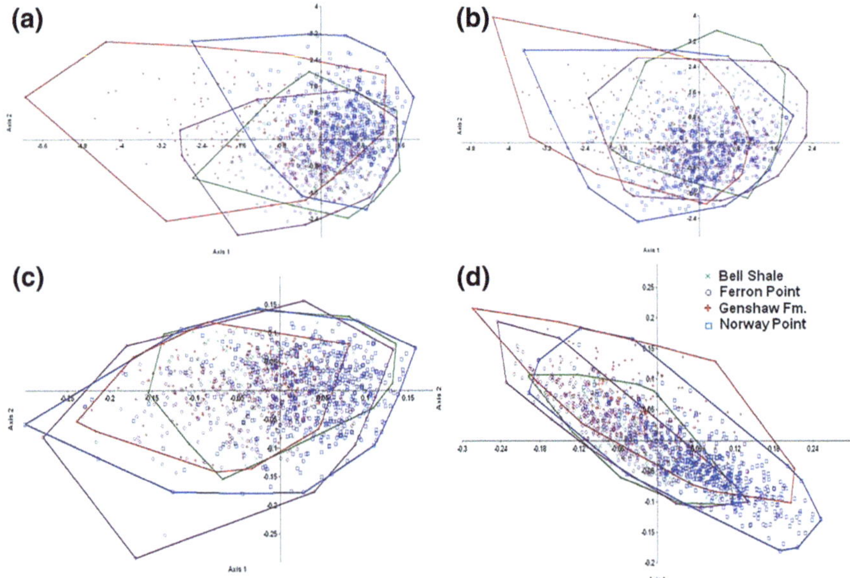

Fig. 3.4 CVA plot showing morphometric differences between samples from Bell Shale, Ferron Point, Genshaw Formation and Norway Point ($p < 0.01$) in **a** dorsal valves, **b** ventral valves, **c** anterior margin area, and **d** posterior hinge area

3.5 Discussion

3.5.1 Patterns of Stasis and Change and Their Possible Causes

Despite the fact that there were significant differences in shell shape between Norway Point and the lower units of the Michigan Basin, the morphological variation within individual units is very large and there is considerable morphological overlap between units. The lower three units show considerable morphological overlap suggestive of stasis-like patterns while the upper Norway Point unit shows abrupt deviation, suggestive of a number of possible causes that are explained later in this section of study.

The morphology of *Pseudoatrypa cf. lineata* was very similar between the three lowermost successive strata (Bell Shale, Ferron Point, Genshaw Formation) with some difference between them and the uppermost strata (Norway Point) (Fig. 3.3). The morphometric shape distances concur with the stratigraphic arrangement of the Traverse Group in that the Euclidean cluster analysis shows that morphological samples from closely spaced strata are more similar than those that are widely spaced in time (Fig. 3.5), thus matching our expectation for gradual change. However, this morphological trend alone is insufficient evidence for determining whether the pattern of change is stasis or gradualism because morphological shape change may be oscillating around a mean. Stronger evidence is required that

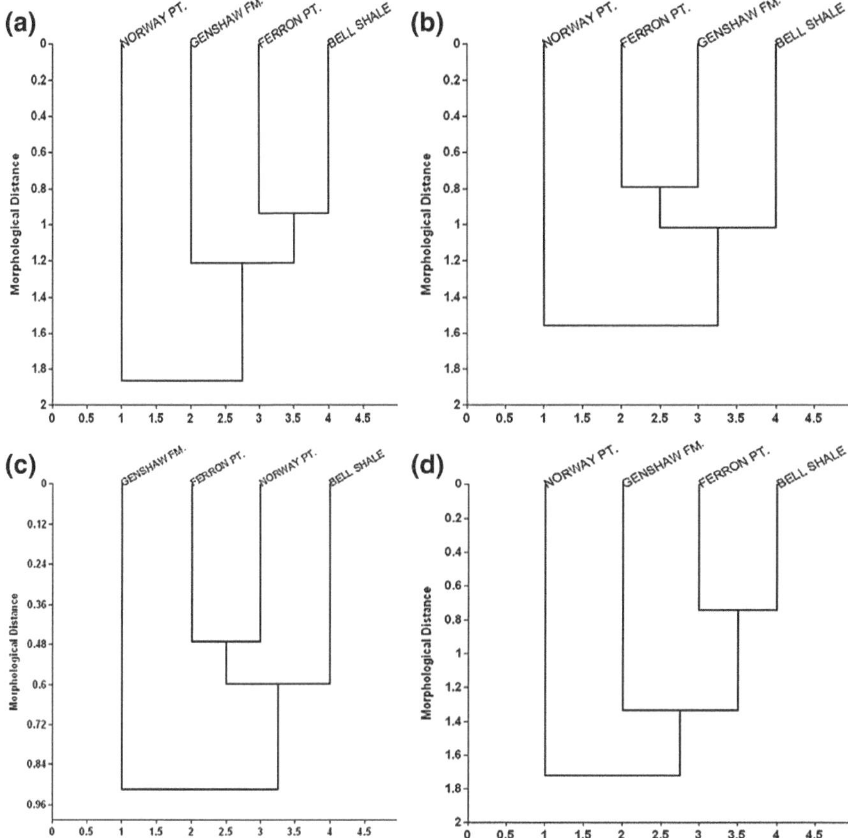

Fig. 3.5 Morphological links for *P. cf. lineata* in Traverse group formations of **a** dorsal valves, **b** ventral valves, **c** anterior margin area, and **d** posterior hinge area

determines the rate and mode of evolution to test whether this shape change pattern is consistent with directional selection. The test based on morphological shape distances plotted against evolutionary divergence time suggests random to static evolution in this taxon with light constraint on morphologies (Fig. 3.6).

Stabilizing selection is one of several processes thought to explain patterns of morphological stasis (Vrba and Eldredge 1984; Maynard Smith et al. 1985; Lieberman et al. 1995; Polly 2004). However, later investigation has shown that a stasis-like pattern may be produced when different selection pressures act on species belonging to different ecosystems, overall, producing no net morphological trend (Lieberman et al. 1994, 1995; Lieberman and Dudgeon 1996). Similarly, when morphological fitness is influenced by many independent environmental variables (e.g., nutrient availability, water depth, storm intensity), morphologies can oscillate in time with changing environments (Polly 2004). Thus, randomly fluctuating selection can explain patterns of morphological change in species over

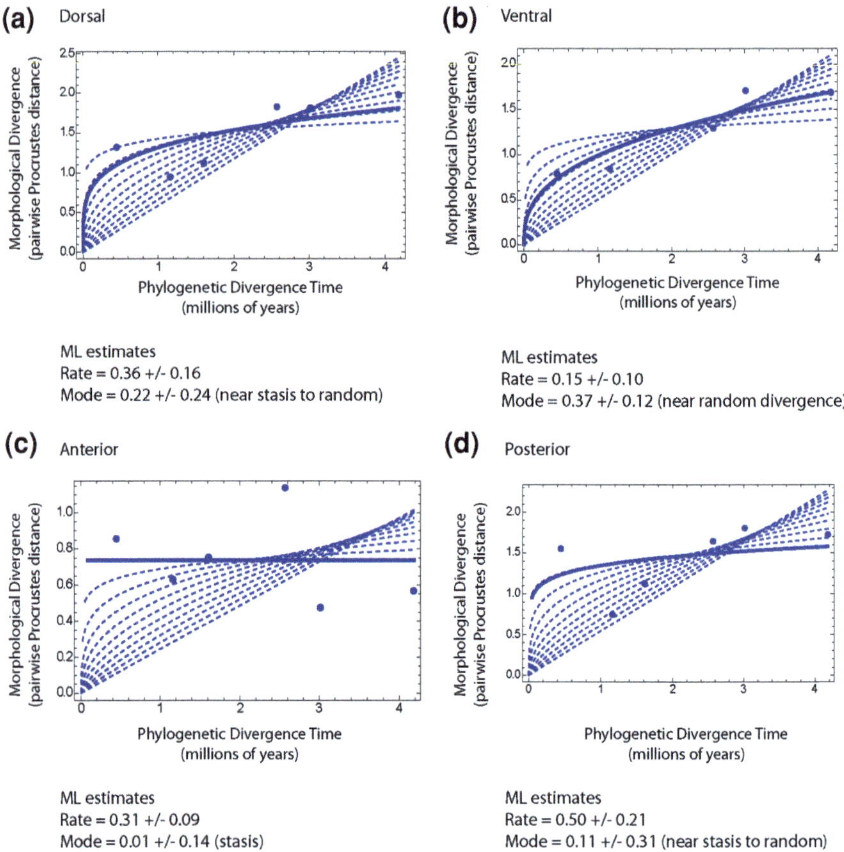

Fig. 3.6 Graph showing morphometric divergence (pairwise Procrustes distances) and phylogenetic divergence time (millions of years). The series of dashed lines show the expected relationship between morphological and phylogenetic divergence time from strong stabilizing selection (0.1), through random divergence (0.5), to diversifying (directional) selection (1.0). The maximum-likelihood estimate of this relationship, shown by the dark line, suggests that *P. cf. lineata* has experienced **a** near stabilizing selection to random divergence in dorsal valves, **b** near random divergence in ventral valves, **c** stabilizing selection in anterior margin area of valves and **d** near stabilizing selection to random divergence in posterior hinge area

time. Here, we have selected samples from more stable environmental regimes with similar lithologic settings, narrow range of water depths, and similar sea level cycles to detect morphological patterns across time. In this study, an evaluation of real morphological distances on brachiopod morphological shape against geologic time suggests that they have evolved predominantly by stabilizing to randomly fluctuating selection. The dorsal valve shows evidence of evolving via near stabilizing to random selection, ventral valve via near random selection, anterior margin via stabilizing selection, and posterior hinge via near stabilizing to random selection. This suggests that morphologies from ancestral to descendant populations in *Pseudoatrypa cf. lineata* evolved statically to randomly, with light constraint on morphologies. The major change in morphology in the upper stratigraphic interval is suggestive of a punctuated change after stable morphologies in the lower stratigraphic intervals.

Four possible causes for the change in the Norway Point sample can be evaluated in this study and are each discussed below. First, this abrupt change may be indicative of either an evolutionary change or a species immigration event, with the Norway Point sample being a different species altogether. Second, this change may have been in response to a somewhat different paleoenvironmental setting represented by the Norway Point interval. Third, the samples from the upper Norway Point may belong to a different EE subunit succession, other than the Hamilton EE subunit. Fourth, the abrupt change may have resulted from a long time interval that elapsed between the lower Genshaw and the Norway Point intervals.

The Norway Point brachiopods were quantitatively different from the Bell Shale, Ferron Point and Genshaw samples. In addition, there seems to be a significant morphological difference between shallow and medium water depth brachiopods. Norway Point brachiopods that preferred shallow water settings were significantly different from those that preferred medium water depths. Thus, this change in morphology may be indicative of an evolutionary change where a new species may have arisen or an immigration event in which another species moved into the Michigan Basin from elsewhere.

Amongst other environmental parameters, water depth has been analysed in correspondence to each stratigraphic interval sampled and it seems that the overall range of water depth for these units varies from <7.0–50.0 m. Notably, Norway Point is shallower (<7.0–8.0 m) than the lower three formations. However, morphology was not correlated with water depth, suggesting that this factor did not cause the observed change. The eustatic curve, in contrast, depicts overall sea level rise punctuated by regression during the final subcycle of Cycle If (Johnson et al. 1985; Wylie and Huntoon 2003), later interpreted as one of the major third order cycles. The Norway Point sequence was deposited during what Brett et al. referred to as the Ih Cycle (Brett et al. 2010). However, sequence stratigraphic data shows that the strata analysed in the Traverse Group represents similar environments throughout with the lower Bell Shale representing the Eifelian-Givetian falling (FSST) sea level, Ferron Point and Genshaw representing Givetian-1 early highstand to falling sea levels (EHST, HST and FSST) and Norway Point representing Givetian-3 highstand (HST) (Brett et al. 2010). Based on detailed correlation records of high order sea level cycles across these basins in Eastern North America, Bell Shale falls within the lower Ie cycle (Johnson et al. 1985), while Ferron Point and Genshaw falls within the If cycle (Brett et al. 2010). Thus, sea level oscillations were probably not responsible for causing this morphological effect either. Lithologically, Norway Point and Ferron Point is made mostly of soft shales with few layers of claystones, Genshaw is made of limestones with shales in the basal part, and Bell Shale is completely shale with a few limestone lenses. Thus, with the samples coming from lithostratigraphically similar settings, it is less likely that this was causing the morphological change. It is however noteworthy, that Norway Point is separated by a thick sequence of limestones below with two erosional unconformities, which is further suggestive of sea level shallowing and the region been subaerially exposed between Genshaw and Norway Point intervals.

Interestingly, environments around the Michigan Basin during the Norway Point time interval show considerable heterogeneity (Dorr and Eschman 1970; Ehlers and Kesling 1970; Wylie and Huntoon 2003) as compared to the other lower stratigraphic intervals (Detroit River Group, Dundee Limestone, lower Traverse formations like the Bell Shale and Alpena Limestone in the Michigan Basin) across different regions (upper, northern, central lower and southeastern lower Peninsula) in the Michigan Basin based on the stratigraphic columns described in Dorr and Eschman (1970). This suggests provinciality in the Michigan Basin during Norway Point time, which raises the possibility that smaller populations may have been isolated in small different local environmental regimes facilitating the greater morphological change observed in the Norway Point sample. Perhaps there were some real environmental changes in the Norway Point interval of the Traverse Group section that may have resulted in environmental selection and thus, leading to evolutionary change.

Samples studied herein are from the Traverse Group of the Michigan Basin that correlates with the Hamilton ecological evolutionary subunit (Bartholomew 2006; Brett et al. 2009, 2010), thus, a good case study to determine morphological patterns in a species lineage from an EESU. However, correlation of the Michigan Basin and Appalachian Basin sections show that Norway Point interval is contemporary with the lower Windom Member of the Moscow Formation of the Hamilton Group based on Orr (1971) and Brett and others (Brett et al. 2010), thus ruling out the possibility for existence of a separate EESU during this time. Though this was the same EESU, the abrupt change observed in Norway Point samples could be due to an immigration event in the Traverse Group caused by transgression after major regression below the Norway Point. Thus, this further suggests that if the Norway Point samples were a different species, then there was a pulse of anagenetic change within *P. cf. lineata* giving rise to a new species (speciation) within the same Hamilton EESU or there was an abrupt change due to species immigration into the Michigan Basin from elsewhere during this time. In either way, the results from this study do not support the ecological locking hypothesis.

Finally, stratigraphic gaps, unconformities and lack of sampling may result in deviation in morphologies. Incorrect recording of spacing between stratigraphic units due to lack of absolute ages and presence of stratigraphic gaps often gives an impression of abrupt change within a species (Sheldon 1987). In this study, samples were collected from units with known differential spacing between them and with almost rarely present stratigraphic gaps between them except for two erosional unconformities observed between Genshaw and Norway Point intervals (Fig. 3.1, Table 3.3). However, the time elapsed in sampling between Genshaw and Norway Point samples, which equals 51.0 m in stratigraphic thickness, is more than that observed between other successive strata (Bell Shale-Ferron Point: 25.0 m, Ferron Point-Genshaw: 35.0 m). Thus, a major change in the Norway Point interval is expected than in any other stratigraphic interval. But it is important to note that the time elapsed between Bell Shale and Genshaw samples, is also nearly 60.0 m; however there is no substantial morphological difference observed between them as observed in the case of Genshaw and Norway Point samples. Thus, the abrupt deflection in morphological variation in Norway Point interval is simply not due to a long interval between Norway Point and other lower strata. In fact, this change may have to do with some kind of anagenetic change in morphological shape pattern as proposed in prior studies (Roopnarine et al. 1999) caused by isolation effects in the Michigan Basin subprovince or due to an immigration event during this time.

The relative importance of stasis has been studied for many fossil lineages (Hunt 2007; Gingerich 2001; Roopnarine 2001; Polly 2001). This study reports for morphological stasis within the *Pseudoatrypa cf. lineata* lineage in the lower Traverse Group units with the advent of abrupt change in the uppermost unit. Though significant statistical differences exist between units, stasis-like patterns evidenced from morphological overlap between lower units followed by change in the uppermost unit samples and results from detailed evolutionary mode analysis, suggest constraint on morphologies in time followed by change. Overall, this is a pattern of punctuated stasis, where there is a long term stasis interrupted by a transitional population later in time. Cases of punctuated stasis have been commonly described from lower Devonian graptolite populations of central Nevada in prior studies (Springer and Murphy 1994). Thus, the pattern observed in this study partially agrees with the punctuated equilibrium model.

3.5.2 Shell Function and Shape Change

Morphological shape change in a species lineage may result from differential functional responses of the various part of the shell during its life. In this study, anterior and posterior morphologies showed more stability relative to other parts of the shell and ventral valves reflect more random change relative to dorsal valves, which showed stasis-like homogeneity. Of the different regions of the shell, the anterior margin was the most stable in contrast to posterior hinge, dorsal and ventral valve area. This anterior margin stability may be related to a functional significance as evident from studies related to frills in atrypids where they are known to stabilize the shells in the substrate (Copper 1967). Other parts reflecting less constraint may be related to their different episodes of life orientation where they were possibly exposed to environmental vagaries during various life stages.

The rate of change also varied with respect to different regions in the shell. The posterior region evolved faster than the dorsal valve, ventral valve and anterior of the shell, while the dorsal valve and anterior evolved even faster than the ventral valve. The posterior region is the hinge of the brachiopods which provides stability to certain life orientations of the animals through pedicle attachment to the substrate during life. During pedicle attachment, immature atrypids are observed in a reclined orientation with their ventral valve facing up and young adults oriented in vertically upright or inclined position, with mature adults oriented with their dorsal valves facing up after pedicle atrophy (Fenton and Fenton 1932; Alexander 1984). Thus, before atrophy, the atrypid brachiopods are attached to the substrate by the pedicles that emerge from the pedicle foramen. The lateral extremes of the hinge line are the regions in which the shell filter feeds on suspended food particles from the host inhalant currents. Thus, this region of the shell must be constantly evolving (a) to resist any strong currents that may be responsible for pedicle atrophy, (b) in response to any encruster settlement on the host that may benefit from host feeding currents acting as parasites, and (c) in response to any predatory attack. The dorsal valve and anterior are probably exposed for greater times during their life in their hydrodynamically stable life position (Fenton and Fenton 1932), thus susceptible to evolving moderately to compensate for any environmental or ecological changes occurring during their life. The ventral valves of the atrypids are facing up during the immature stages of their life (Alexander 1984) with probably less exposure time, thus susceptible to fewer episodes of selection, which further explains the slow rate of evolution of this region as compared to other regions in the shell.

3.6 Conclusion

Landmark measurements in atrypid species lineage *Pseudoatrypa cf. lineata* from the Middle Devonian Traverse Group of Michigan State has been analysed in this study to determine whether morphological shape trends in a lineage can be explained by punctuated equilibrium model. Geometric morphometric and multivariate statistical analyses reveals significant statistical differences in morphological shape between Traverse group stratigraphic units with considerable overlap in widespread morphologies over 5 m.y. interval of time. Notably, the samples from the uppermost strata, Norway Point formation show an abrupt morphological shift from the lower stratigraphic units, Bell Shale, Ferron Point, and Genshaw Formation. Thus, this suggests that morphological shape, a species diagnostic character, underwent very little change in the lower Traverse Group stratigraphic intervals with major change been reflected in the upper Norway Point formation. This change may be attributed to the environmental heterogeneity observed during this time with two erosional unconformities below the Norway Point, further leading to high provinciality in the Michigan Basin sections during the Norway Point time interval. Thus, two possible explanations for this abrupt change may be real evolutionary change due to environmental selection or new species immigration to the Michigan Basin from elsewhere.

The maximum-likelihood method suggests a slow to moderate rate of evolution with near stasis to random divergence mode of evolution in the *Pseudoatrypa cf.*

lineata species lineage. However, different parts of the shell show light constraint on morphologies with anterior and posterior showing greater constraint than ventral valves and ventral valves even more constraint than dorsal valves. Evolutionary stasis in anterior region may be a response of the stabilizing function of the frills in their anterior margin. Ventral valves evolving at a lesser magnitude than all other shape measurements (dorsal valves, posterior hinge area and anterior margin), is indicative of its being subject to fewer episodes of selection as a result of lower residence time in its ventral valve facing up early stage life orientation. While ventral valves show maximum fluctuation in their evolutionary mode, anterior margin is most stable in morphology over time.

Overall, the mean shape morphological trend suggests considerable morphological overlap between the lower successive stratigraphic units of the lower Traverse Group with small morphological oscillations in the species life history. However, samples in the uppermost strata deviate from the mean so far, such that either anagenetic evolution may have caused the rise of new descendant populations or new species from elsewhere may have replaced the native species later in time.

Morphology of *Pseudoatrypa cf. lineata* weakly correlates with variation in water depth. Samples from shallow versus medium water depths show significant difference in morphology. This difference in morphology could be attributed to the difference in species, one of which preferred shallow water depths and others which preferred medium water depths. Since testing other environmental variables is beyond the scope of this study, it was challenging to infer if any abiotic or biotic factors were behind these mechanisms of evolutionary selection. Thus, this suggests, that the morphological shape trend can be explained only by stabilizing selection and/or by randomly fluctuating selection mechanisms trigerred by environmental changes occurring in that regime.

Thus, the results from this study appear to partially comply with the punctuated equilibrium model as morphological shape response in *Pseudoatrypa cf. lineata* in the Traverse Group of the Michigan Basin is more in agreement with near stabilizing to random selection processes with constraint on morphologies rather than directional selection process. Stasis is represented in the *P. cf. lineata* species with evidence of morphological change later in time. This abrupt change may be attributed to either an evolutionary change resulting from environmental selection or an immigration event that occurred during that time.

References

Alexander RR (1984) Comparative hydrodynamic stability of brachiopod shells on current-scoured arenaceous substrates. Lethaia 17:17–32

Bookstein FL (1991) Morphometric tools for landmark data: geometry and biology. Cambridge University Press, Cambridge, p 435

Bartholomew AJ (2006) Middle devonian faunas of the Michigan and Appalachian Basins: comparing patterns of biotic stability and turnover between two paleobiogeographic subprovinces, Unpublished Ph.D. dissertation, p 300

Bartholomew AJ, Brett CE (2007) Correlation of Middle Devonian Hamilton Group-equivalent strata in east-central North America: implications for eustasy, tectonics and faunal provinciality: geological society, vol 278. Special Publications, London, pp 105–131

Brett CE (1986) The Middle Devonian Hamilton group of New York: an overview. In: Brett C (ed) Dynamic stratigraphy and depositional environments of the Hamilton Group (Middle Devonian) in New York State, Part 1: New York state museum bulletin, vol 457, pp 1–4

Brett CE, Baird GC (1995) Coordinated stasis and evolutionary ecology of silurian to Middle Devonian faunas in the Appalachian Basin. In: Erwin DH, Anstey RL (eds) New approaches to speciation. Columbia University Press, NewYork, pp 285–315

Brett CE, Ivany LC, Bartholomew AJ, DeSantis MK, Baird GC (2009) Devonian ecological-evolutionary subunits in the Appalachian Basin: a revision and a test of persistence and discreteness: geological society, vol 314. Special Publications, London, pp 7–36

Brett CE, Baird GC, Bartholomew AJ, DeSantis MK, Straeten CA (2011) Sequence stratigraphy and a revised sea-level curve for the Middle Devonian of eastern North America. Palaeogeogr Palaeoclimatol Palaeoecol 304:21–53

Carlson EH (1991) Minerals of Ohio: division of geological survey bulletin, vol 69, p 155

Cooper GA, Butts C, Caster KE, Chadwick GH, Goldring W, Kindle EM, Kirk E, Merriam CW, Swartz FM, Warren PS, Warthin AS, Willard B (1942) Correlation of the Devonian sedimentary formations of North America. Geol Soc Am Bull 53:1729–1794

Cooper G, Dutro JT (1982) Devonian brachiopods of New Mexico. Bull Amer Paleont 82–83(313):215

Copper P (1967) Adaptations and life habits of Devonian atrypid brachiopods. Palaeogeogr Palaeoclimatol Palaeoecol 3:363–379

Copper P (1973) New Siluro-Devonian atrypoid brachiopods. J Paleontol 47:484–500

Day J (1992) Middle-Upper Devonian (Late Givetian–Early Frasnian) brachiopod sequence in the Cedar Valley Group of central and eastern Iowa. In: Day J, Bunker BJ (eds) The stratigraphy, paleontology, depositional and diagenetic history of the Middle-Upper Devonian Cedar Valley Group of Central and Eastern Iowa: Iowa department of natural resources guidebook series, vol 16, pp 53–105

Day J (1996) Faunal signatures of Middle-Upper Devonian depositional sequences and sea level fluctuations in the Iowa Basin: U.S. midcontinent. In: Witzke BJ, Ludvigson GA, Day J (eds) Paleozoic sequence stratigraphy, views from the North American craton. Geological society of America special paper, vol 306, pp 277–300

Day J, Copper P (1998) Revision of latest Givetian-Frasnian Atrypida (Brachiopoda) from central North America. Acta Palaeontologica Polonica 43:155–204

Dorr JA, Eschman DF (1970) Geology of Michigan. The University of Michigan Press, Ann Arbor, p 479

Ehlers GM, Kesling RV (1970) Devonian strata of Alpena and Presque Isle counties, Michigan. Michigan basin geological society, p 130

Eldredge N, Gould SJ (1972) Punctuated equilibria. In: Schopf TJM (ed) Models in paleobiology. Freeman, Cooper and company, San Francisco, pp 82–115

Ettensohn FR (1985) The Catskill Delta complex and the Acadian orogeny: a model. In: Woodrow DL, Sevon WD (eds) The Catskill Delta, special paper—geological society of America, Boulder, vol 201, pp 39–50

Felsenstein J (1988) Phylogenies and quantitative characters. Annu Rev Ecol Syst 19:445–471

Fenton CL, Fenton MA (1932) Orientation and Injury in the Genus *Atrypa*. Am Midl Nat 13:63–74

Fenton CL, Fenton MA (1935) Atrypae described by Clement L. Webster and related forms (Devonian, Iowa). J Paleontol 9:369–384

Gingerich PD (1993) Quantification and comparison of evolutionary rates. Am J Sci 293A:453–478

Gingerich PD (2001) Rates of evolution on the time scale of the evolutionary process. Genetica 112–113:127–144

Goldman D, Mitchell CE (1990) Morphology, systematics, and evolution of Middle Devonian Ambocoeliidae (Brachiopoda), Western New York. J Paleontol 64:79–99

Gould SJ, Eldredge N (1977) Punctuated equilibria: the tempo and mode of evolution reconsidered. Paleobiology 3:115–151

Hunt G (2007) The relative importance of directional change, random walks, and stasis in the evolution of fossil lineages. Proc Natl Acad Sci 104:18404–18408

Isaacson PE, Perry DG (1977) Biogeography and Morphological Conservatism of *Tropidoleptus* (Brachiopoda, Orthida) during the Devonian. J Paleontol 51:1108–1122

Johnson JG, Klapper G, Sandberg CA (1985) Devonian eustatic fluctuations in Euramerica. Geol Soc Am Bull 96:567–587

Kesling RV, Segall RT, Sorensen HO (1974) Devonian strata of Emmet and Charlevoix counties, Michigan. Michigan Museum Paleontol Papers Paleontol 7:1–187

Koch WF (1978) Brachiopod paleoecology, paleobiogeography, and biostratigraphy in the upper middle Devonian of Eastern North America: an ecofacies model for the Appalachian, Michigan, and Illinois basins. Unpublished Ph.D. Thesis, Oregon State University, Oregon, p 311

Lande R (1976) Natural selection and random genetic drift in phenotypic evolution. Evolution 30:314–334

Lieberman BS, Brett CE, Eldredge N (1994) Patterns and processes of stasis in two species lineages from the Middle Devonian of New York State. AMNH 3114:1–23

Lieberman BS, Brett CE, Eldredge N (1995) A study of stasis in two species lineages from the Middle Devonian of New York State. Paleobiology 21:15–27

Lieberman BS, Dudgeon S (1996) An evaluation of stabilizing selection as a mechanism for stasis. Palaeogeogr Palaeoclimatol Palaeoecol 127:229–238

Maynard Smith J, Burian R, Kauffman S, Alberch P, Campbell J, Goodwin B, Lande R, Raup D, Wolpert L (1985) Developmental constraints and evolution. Q Rev Biol 60:265–287

Morris PJ (1995) Coordinated stasis and ecological locking. Palaios 10:101–102

Morris PJ, Ivany LC, Schopf KM, Brett CE (1995) The challenge of paleoecological stasis: reassessing sources of evolutionary stability. Proc Natl Acad Sci 92:11269–11273

Orr W (1971) Conodonts from Middle Devonian Strata of the Michigan Basin. Indiana Geological Survey Bulletin, vol 45, p 110

Polly PD (2001) Paleontology and the comparative method: ancestral node reconstructions versus observed node values. Am Nat 157:596–609

Polly PD (2004) On the simulation of morphological shape: mutivariate shape under selection and drift. Palaeontologia Electronica 7(7A):1–280

Polly PD (2008) Adaptive zones and the pinniped ankle: a three-dimensional quantitative analysis of carnivoran tarsal evolution. Mammalian evolutionary morphology, pp 167–196

Rohlf FJ (1990) Fitting curves to outlines. In: Proceedings of the Michigan morphometrics workshop, pp 167–177

Rohlf FJ, Slice DE (1990) Extensions of the procrustes method for the optimal superimposition of landmarks. Syst Zool 39:40–59

Rohlf FJ (1999) Shape statistics: procrustes superimpositions and tangent spaces. J Classif 16:197–223

Rohlf FJ, Marcus LF (1993) A revolution in morphometrics. Trends Ecol Evol 8:129–132

Roopnarine PD, Byars G, Fitzgerald P (1999) Anagenetic evolution, stratophenetic patterns, and random walk models. Paleobiology 25:41–57

Roopnarine PD (2001) The description and classification of evolutionary mode: a computational approach. Paleobiology 27:446–465

Roopnarine PD (2003) Analysis of rates of morphologic evolution. Annu Rev Ecol Evol Systs 34:605–632

Sheldon PR (1987) Parallel gradualistic evolution of ordovician trilobites. Nature 330:561–563

Slice DE (2001) Landmark coordinates aligned by procrustes analysis do not lie in kendall's shape space. Syst Biol 50:141–149

Springer KB, Murphy MA (1994) Punctuated stasis and collateral evolution in the Devonian lineage of *Monograptus hercynicus*. Lethaia 27:119–128
Stanley SM (1979) Macroevolution. W.H. Freeman, San Francisco
Stumm EC (1951) Check list of fossil invertebrates described from the middle Devonian Traverse group of Michigan: contributions from the museum of paleontology, University of Michigan vol 9, pp 1–44
Vrba ES, Eldredge N (1984) Individuals, hierarchies and processes: towards a more complete evolutionary theory. Paleobiology 10:146–171
Webster CL (1921) Notes on the genus *Atrypa*, with description of new species. Am Midl Nat 7:13–26
Wiedman LA (1985) Community paleoecological study of the silica shale equivalent of northeastern Indiana. J Paleontol 59:160–182
Wylie AS, Huntoon JE (2003) Log-curve amplitude slicing: visualization of log data and depositional trends in the Middle Devonian Traverse Group, Michigan Basin, USA

Chapter 4
Morphological Shape, Episkeletobiont Analysis, and Life Orientation Study in *Pseudoatrypa cf. lineata* (Brachiopoda) from the Lower Genshaw Formation of the Middle Devonian Traverse Group, Michigan: A Geometric Morphometric Approach

4.1 Introduction

Pseudoatrypa is a common brachiopod from the Givetian to late Frasnian of North America. This genus occurs throughout much of the Traverse Group in the Michigan Basin, including the Genshaw Formation (Kelly and Smith 1947; Koch 1978). Here we focus on material from the lower Genshaw Formation to: (1) analyze morphological shape patterns in two *Pseudoatrypa* species, and, (2) investigate episkeletobiont interactions with these species to determine how the distinct morphological shapes of the two species may have influenced their settlement.

Pseudoatrypa is frequently encountered in Devonian Midcontinent basins. Webster (1921) first described the taxon as *Atrypa devoniana* from the late Frasnian Independence Shale of Iowa; his specimens were later designated as the type species of the new genus *Pseudoatrypa* (Copper 1973; Day and Copper 1998). *Pseudoatrypa* also occurs in the Traverse Group of the Michigan Basin (Stumm 1951; Copper 1973; Koch 1978) and in the Silica Formation of Ohio and equivalent rocks from northern Indiana (Wiedman 1985). Webster (1921) described the species *Atrypa lineata* from the late Middle Devonian (late Givetian) upper Osage Springs Member of the Lithograph City Formation of the Upper Cedar Valley Group of Iowa, which was later included in *Pseudoatrypa* by Day and Copper (1998). Fenton and Fenton (1935) described a subspecies and growth variant forms of this species from the late Givetian Cedar Valley of Illinois. Herein we test both qualitatively and quantitatively whether distinct species exist within *Pseudoatrypa* from the Lower Genshaw Formation of Michigan, and whether these external morphology influences episkeletobiont assemblages.

Episkeletobionts are organisms that adhere to, or encrust, the surface of a shell (Taylor and Wilson 2002). Episkeletobionts are useful as ecological and life status indicators of their hosts—whether the host was living at the time of encrustation or

R. Bose, *Biodiversity and Evolutionary Ecology of Extinct Organisms*,
Springer Theses, DOI: 10.1007/978-3-642-31721-7_4,

was dead (Watkins 1981; Anderson and Megivern 1982; Brezinski 1984; Gibson 1992; Lescinsky 1995; Sandy 1996; Sumrall 2000; Morris and Felton 2003; Schneider 2003, 2009a; Zhan and Vinn 2007; Rodrigues et al. 2008). Episkeletobionts have been used to infer the life orientation of brachiopods (Rudwick, 1962; Hurst 1974; Pitrat and Rogers 1978; Kesling et al. 1980; Spjeldnaes 1984; Lescinsky 1995), the preferred orientation of host water currents (e.g., Kesling et al. 1980), potential camouflage for hosts (Schneider 2003, 2009a), the attracting or antifouling nature of ornamentation (Richards and Shabica 1969; Richards 1972; Carrera 2000; Schneider 2003, 2009a; Schneider and Leighton 2007), and the function of the valve punctae (Thayer 1974; Curry 1983; Bordeaux and Brett 1990). Brachiopod hosts are useful for investigating host influences on episkeletobiont preferences such as shell texture (Schneider and Webb 2004; Rodland et al. 2004; Schneider and Leighton 2007), size of host (Ager 1961; Kesling et al. 1980), and antifouling strategies (Schneider and Leighton 2007). Although other Paleozoic marine organisms were frequently encrusted, brachiopods remain one of the best understood hosts for Paleozoic episkeletobionts.

In the present study, the settlement of a live episkeletobiont during the life of the host is called a live–live association (pre-mortem encrustation) and the settlement of a live encruster on a dead host is called a live-dead association (post-mortem encrustation). Taylor and Wilson (2003) provided the following criteria for distinguishing between pre- and post-mortem associations: (a) If the episkeletobiont fossil overgrows (crosses) the commissural margin, or if there is evidence of internal valve encrustation, then the brachiopod host was dead when the organism overgrew the commissure—this is evidence of a live-dead association; (b) if the episkeletobiont and the host both have a similar degree of preservation, and if there is evidence of scars representing the repair of damage inflicted by episkeletobionts, then the brachiopod host was alive during encrustation—this is evidence of live–live association; and (c) if certain episkeletobionts repeatedly encrust specific locations on host shells, e.g., if branching fossils, like auloporid corals or hederellids, branch towards or are aligned parallel to the commissure or if solitary organisms, such as cornulitids, grow with their apertures pointing towards the commissure, then hosts and their episkeletobionts experienced live–live associations. In other cases, there is no way to tell for certain whether the host was alive or dead at the time of encrustation.

Our purpose herein is twofold: (1) to quantitatively assess putatively distinct species of *Pseudoatrypa* from the Genshaw Formation, herein described as *P. lineata* and *Pseudoatrypa* sp. A; and (2) to examine the influence of species morphology on encrustation by episkeletobionts. We structured the study by testing the following hypotheses:

(1) two species of *Pseudoatrypa*—*P. lineata* and *Pseudoatrypa* sp. A, previously distinguished on qualitative features, must have statistically distinct shell shapes and are validly different species;if not they will be considered growth variants of the same species;

(2) episkeletobionts are influenced by morphology and will preferentially encrust the two species differently; and

(3) given the preferred orientation of atrypid adult shells—convex dorsal valve raised into the water column and flat ventral valve in contact with the substrate (as inferred by Fenton and Fenton (1932)) —the extent of encrustation coverage on the ventral valve will be limited by physical contact with the substrate and thus statistically less than on the dorsal valve.

These hypotheses were tested using geometric morphometric assessment of shell shape and statistical analysis of the location of encrusting organisms.

4.2 Geologic Setting

Pseudoatrypa brachiopods were collected from the lower Genshaw Formation of the Middle Devonian Traverse Group. The Traverse Group ranges in thickness from ∼25.0–169.5 m (Ehlers and Kesling 1970; Wylie and Huntoon 2003), with depositional environments ranging from shallow water carbonate lagoons with coral-stromatoporoid reefs to storm-dominated mixed carbonate-siliciclastic shelf deposits and offshore muddy shelf to slope environments (Ehlers and Kesling 1970). The Genshaw Formation was named by Warthin and Cooper (1935) for strata overlying their Ferron Point Formation and underlying their Killians Limestone, later revised by the same stratigraphers to include the Killians Lime-stone as the upper member of the Genshaw Formation (Warthin and Cooper 1943). Warthin and Cooper (1943) placed the new upper contact at the base of the overlying Newton Creek Limestone. The Genshaw Formation remained one of the least studied units of the lower Traverse Group until recently, when the LaFarge Quarry in the Alpena area began to mine into this unit and exposed nearly the entire Formation (Bartholomew 2006). The Genshaw Formation accumulated during the highstand of a third-order sea level sequence (Wylie and Huntoon 2003; Brett et al. 2010).

The Genshaw Formation, which is ∼30.0 m thick (Fig. 4.1), is subdivided into informal lower, middle, and upper (formerly Killians Member) portions (Wylie and Huntoon 2003). The lower unit of the Genshaw Formation begins with a 0.5 m-thick crinoidal grainstone, which locally contains burrows on its lower surface. Overlying this basal bed of the Genshaw Formation is a thin, argillaceous succession capped by a limestone-rich interval to the top of the lower Genshaw. The brachiopods used in this study were collected from the argillaceous beds of this lower unit (Fig. 4.1).

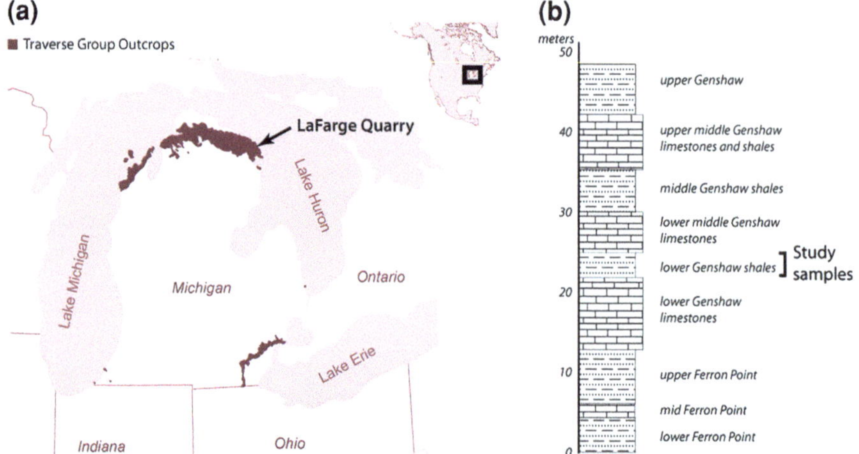

Fig. 4.1 **a** Map showing Michigan surface exposures of the Middle Devonian Traverse Group and the location of the La Farge Quarry, Alpena area, from which the samples used in this study were collected, **b** Simplified stratigraphic section of the Traverse Group at La Farge Quarry showing the horizons of the Genshaw Formation where the specimens used in this study were collected (after Bartholomew 2006)

4.3 Materials and Methods

4.3.1 Sampling

Samples examined were collected by A. Bartholomew of State University of New York, New Paltz from the northeastern outcrop of the Lafarge Alpena Quarry, Alpena County, Michigan (Fig. 4.1). He extensively sampled all brachiopods from a shale bed of the lower Genshaw Formation. The 185 well-preserved atrypids examined for encrustation in this study have been deposited in the Indiana University Paleontology Collection (IU 100059–IU 100243). Use of ammonium chloride spray in a dry environment helped distinguish morphological features of the species.

4.3.2 Species Recognition

The atrypid sample was first divided into two populations based on the qualitative traits examined in this study. The two populations are similar in that they have an apical foramen, hinge line with incurved extremities, orbicular to subquadrate shell outline, ribbing with implantations and bifurcations, and somewhat similar spacing between growth lamellae or frills, with frills crowding more at the anterior. However, *P. lineata* is different from *Pseudoatrypa* sp. A in having (a) a smooth, arcuate, domal curvature to the dorsal valve as opposed to the arched shape in the latter, (b) slightly inflated ventral valve with an inflation near the umbo as opposed to a more flattened ventral valve in the latter, (c) relatively lower dorsal valve curvature height, (d) fine to medium closely spaced ribs in contrast to coarse ribs in the latter, (e) angular to subrounded hinge line as opposed to the widened hinge in the latter, (f) widened median deflection (fold and sulcus) on the commissure as opposed to the narrow deflection in the latter, and (g) gentle to steep mid-anterior fold as opposed to the gentler fold in the latter.

4.3.3 Morphometrics

We performed geometric morphometric analysis to determine morphological variation within and between the two qualitatively distinguished populations assigned to *P. lineata* and *Pseudoatrypa* sp. A. No crushed specimens were used for morphometrics.

Geometric morphometrics is the analysis of geometric landmark coordinates points on specific parts of an organism (Bookstein 1991; MacLeod and Forey 2002; Zelditch et al. 2004). We based our morphometric analysis on the use of landmarks to capture shape (Rohlf and Marcus 1993); landmarks represent discrete geometric points on each specimen that correspond among forms (sensu Bookstein 1991). In this study, we used 10 two-dimensional landmark points on the external shell to capture the most meaningful shape differences (Fig. 4.2). When selecting landmarks for analyses, we selected points that not only characterized body shape accurately, but also represented some aspect of the inferred ecological niche. All landmarks were defined by geometric position on host shells (1 = beak tip on brachial valve; 2 and 8 = intersection points of the commissure and the hingeline; also region for food intake from inhalant currents; 3 and 7 = length midpoint projected onto the commissure; these points are perpendicular to, and crosses the midline of the shell; 4 and 6 = lowest point of median deflection (fold and sulcus) on the commissure; 5 = middle point of commissure; 9 = tip of umbo on pedicle valve; 10 = maximum height on brachial valve). These landmarks are appropriate for analyses attempting to capture shape changes or function. For this study, we conducted four different analyses operating on ten landmarks in four different orientations of the shell. Two analyses were conducted in the x–y plane of the dorsal and ventral valves; these analyses capture only the view in that plane (Fig. 4.2a, b). These analyses of the dorsal and ventral valve views, included nine landmarks, which were selected to encompass the outline of the entire specimen in the x–y plane. Although brachiopods are bilaterally symmetrical, landmarks were included from both the left and right sides of the specimens to record the functional response of these hosts to the then existing ecological conditions and to encrusting episkeletobionts. Capturing both the postero-lateral distal extremities of the hingeline and the lowest points of median deflection on the commissure, even of a bilaterally symmetrical organism may be important for determining shape

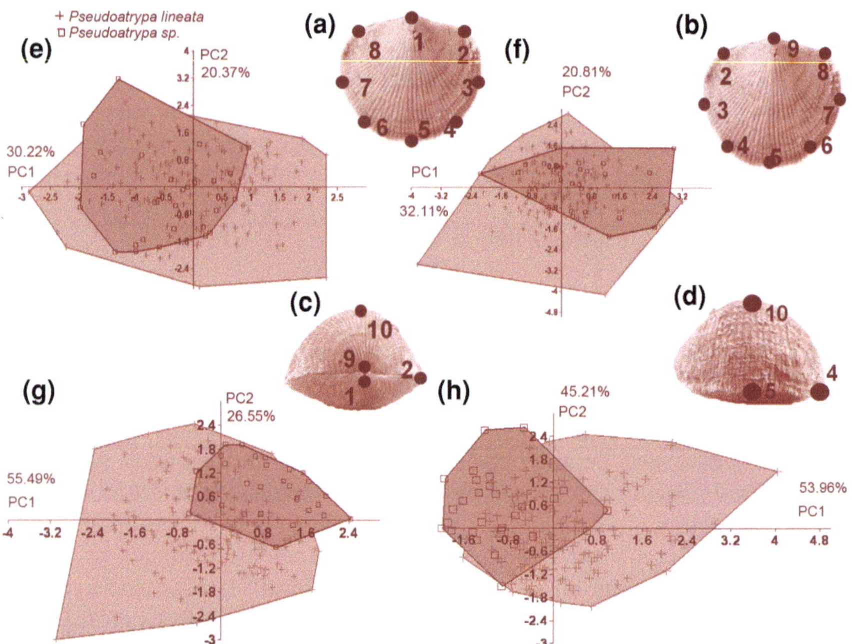

Fig. 4.2 **a** Location of eight landmark points on dorsal valves of host species, **b** location of eight landmark points on ventral valves of host species, **c** location of four landmark points on posterior region of host species, **d** location of three landmark points on anterior region of host species; morphological shape variation between the two host species *P. lineata* and *Pseudoatrypa* sp. A along **e** dorsal vales, **f** ventral valves, **g** posterior region, and **h** anterior region

changes in the host species, as each of these locations may possess unique specific abundances of distinct episkeletobiont assemblages (Bookstein 1991; Kesling et al. 1980). Two separate analyses operating on landmarks in the y–z plane were conducted from the anterior and posterior views of the shells (Fig. 4.2c, d). For posterior and anterior regions of the shell, landmark measurements (four landmarks on posterior and three landmarks on anterior) were taken only on half of the specimen (anterior/posterior view left or right) (Fig. 4.2c, d). Overall, these two orientations measure not only the shape of the valves, but also capture the shape of the brachiopod lophophore support, the spiralia (Bookstein 1991; Haney et al. 2001). The four views (dorsal, vental, anterior, and posterior) were analyzed separately.

Procrustes analysis (Rohlf 1990; 1999; Rohlf and Slice 1990; Slice 2001) was performed on original shape data, rotating, translating and scaling all landmarks to remove all size effects, while maintaining their geometric relationships (Procrustes superimposition). Principal component analysis of the covariance matrix of the residuals of the Procrustes superimposed coordinates was performed to determine the morphological variation of the two species along their major principal component axes (1 and 2) in the shape morphospace (Fig. 4.2) and to provide a set of uncorrelated shape variables for further statistical analysis. Procrustes distances, which are the sum of the distances between corresponding landmarks of Procrustes superimposed objects (equal to their Euclidean distance in the principal components space if all axes are used), were calculated as a measure of difference between mean morphological shapes of the dorsal valves, ventral valves, posterior and anterior regions of *P. lineata* and *Pseudoatrypa* sp. A and thin plate spline plots (Bookstein 1989) were used to visualize those differences. Multivariate analysis of variance (MANOVA) and discriminant function analysis (DFA) of the shape variables was used to test for significant shell shape differences between the two species along the dorsal, ventral, anterior and posterior regions of the shells (Hammer and Harper 2005). Multivariate analysis of variance (MANOVA) was performed to test shape variation between the two species. Adult shells of the two species, larger than 1.9 cm in size, were used for geometric morphometric analyses to avoid any misinterpretation in comparative shape study between the two species that could have resulted from not controlling for ontogenic development.

4.3.4 Episkeletobiont Analysis

Brachiopods were first examined microscopically for encrustation data (100x magnification). Episkeletobiont distribution was tabulated for each species, for valve preference within each species, and for position on each valve. The relationship between these episkeletobionts and the host atrypids were investigated based on their placement on the host shell. The *Chi-square* test was used to determine whether differences in most abundant episkeletobiont assemblages were significant between the two host species and between the dorsal and ventral valves of each species. Furthermore, species and valve preference by episkeletobionts was reconstructed based on the abundance and location of episkeletobionts, as described below.

Mean encrustation frequency per species was determined to compare encrustation abundance on the dorsal and ventral valves of the two species as

$$A_C = \left[\left(\sum E_T / \sum V_E \right) / N \right] * 100, \tag{1}$$

where A_C is the mean encrustation frequency with respect to episkeletobiont count, E_T is the total number of episkeletobiont colonies, and V_E is the total number of valves encrusted.

The total area of the valve that was encrusted was measured on each individual host, and the proportion of the valve that was encrusted was calculated using the equation:

$$A_A = A_E / A_V * 100, \tag{2}$$

where A_A is the encrustation area per valve (A_E) with respect to total valve area (A_V).

Relationship between encrustation area (A_{A-}) and principal component (PC axis 1) scores of the shape co-ordinates of the valve morphology was tested using the product-moment correlation (r). A *p value* was also reported for this correlation method to determine if 'r' was significantly different from 0.0.

Following the methods previously established by Bose et al. (2010), each valve of the atrypid hosts was divided into six regions (Fig. 4.3). The six regions were defined as postero-left lateral (PLL), posteromedial (PM), postero-right lateral (PRL), antero-left lateral (ALL), anteromedial (AM), and antero-right lateral (ARL). These six divisions, i.e., six different surface areas of host, were selected such that they represent biologically functional grids for both the host and the episkeletobiont. The PLL and PRL regions were selected based on the idea that host inhalant currents in those regions may attract episkeletobionts, and these currents may also partially influence episkeletobiont settling along the PM region. Similarly, the AM region was selected based on the host exhalant current criteria, which may also partially influence the ALL and ARL regions. Thus, selecting these six regions and recording the frequency of episkeletobionts on each of these regions will help infer host-episkeletobiont relationships in live–live associations.

Area ratios for each region of the shell was determined using the following equation (Fig. 4.3):

$$\mathbf{R} = A_R / A_T \tag{3}$$

where R is the area ratio, A_R is the area of each region, and A_T is the total area.

The frequency of encrustation in each region was recorded by counting individual colonies as one occurrence and then summing for all atrypid hosts. For comparison with the actual frequency of episkeletobiosis, expected episkeletobiosis for each region was calculated by:

$$\mathbf{E} = NR_i \tag{4}$$

in which the expected number of episkeletobionts (E) is calculated by multiplying the total number of episkeletobionts (N) for all *Pseudoatrypa* specimens by the area ratio (R) for a given region on the atrypid shell (i). The null hypothesis used here for developing the expected value is as follows: Given that we have six regions for one valve, and each region has a chance of being encrusted based on their proportion of available surface area (assuming as a null hypothesis a random distribution of encrusters), then if one region accounts for x % of the surface area on the valve, then the expected value for that region is x % of the total number of episkeletobionts encrusting that valve (all six regions) for that species. The same approach is used to determine the expected value for the other regions. Colonial episkeletobionts—sheet-like and branching—posed a problem for calculations because these specimens often crossed borders into adjacent regions. For these specimens, colonization of an episkeletobiont that extended into two or more regions was divided among the total number of regions it inhabited. For example,

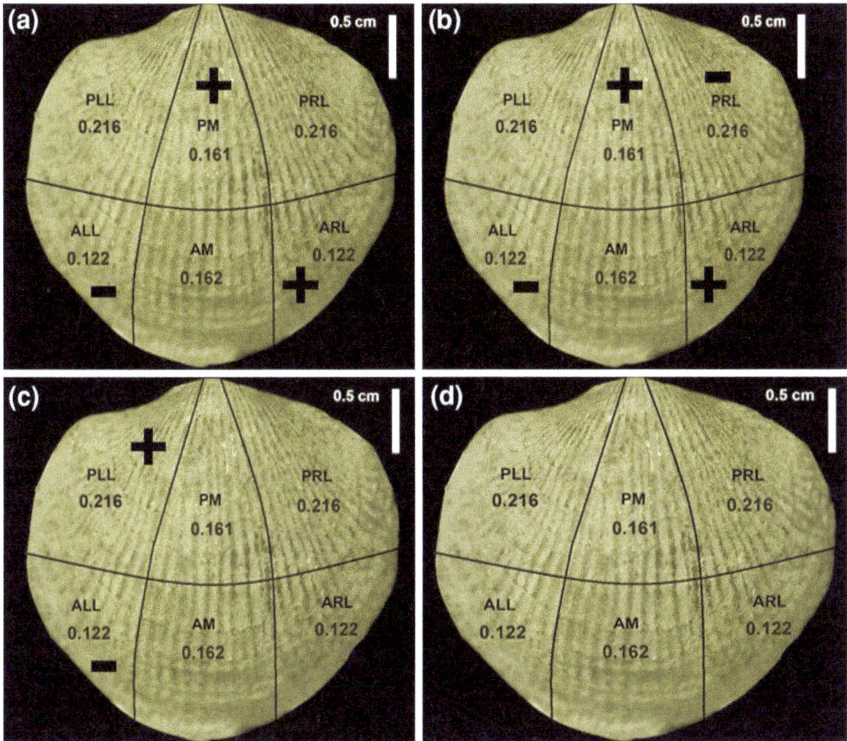

Fig. 4.3 *Pseudoatrypa* specimen divided into six regions for episkeletobiont frequency study; *PLL* postero-left lateral, *PM* posteromedial; *PRL* postero-right lateral; *ALL* antero-left lateral, *AM* anteromedial; and *ARL* antero-right lateral. Numbers represent the area ratios of each grid across the *P. lineata* and *Pseudoatrypa* sp. A host valve. Scale bar 0.5 cm. Note that the area ratios are slightly different for *P. lineata* host species which are as follows: PLL = 0.18, PM = 0.138, PRL = 0.18, ALL = 0.147, AM = 0.208, ARL = 0.147. Significantly greater observed episkeletobiont frequency than expected is denoted by a *plus symbol* and smaller observed episkeletobiont frequency than expected is denoted by a *minus symbol* across the six regions of the valve; this is described for **a** dorsal valve of *P. lineata*, **b** dorsal valve of *Pseudoatrypa* sp. A, **c** ventral valve of *P. lineata*, and **d** ventral valve of *Pseudoatrypa* sp. A

branching auloporid corals and hederellids that were observed in all the three anterior regions were counted as 1/3 for each region.

We then quantified common episkeletobionts (i.e., microconchids, bryozoans sheets, and hederellids) for their distribution on six regions of the shell of both dorsal and ventral valves of each host, using Equation 2 above. Distribution of rare episkeletobionts on host shells was also examined, but only for dorsal valves, as episkeletobiont abundance of rare encrusters is negligible on ventral valves. A *Chi-square* test was also performed for the total observed and expected episkeletobiont activity along the six regions to determine the episkeletobiont location preference on the atrypid valve within each species.

Fig. 4.4 *Pseudoatrypa lineata* **a** Dorsal view, **b** ventral view, **c** posterior view, **d** anterior view (IU#100069); *Pseudoatrypa* sp. A **e** dorsal view, **f** ventral view, **g** posterior view and **h** anterior view (IU#100220). The small inset illustrations next to dorsal and ventral views of *P. lineata* represent the type specimen of *Atrypa lineata var. inflata* as described in Fenton and Fenton 1935 and the posterior and anterior views represent the type specimen of *Atrypa lineata* as described in Day and Copper 1998

4.4 Results

4.4.1 Morphology and Morphometrics

Two-hundred and thirty two specimens of atrypids were assigned to *P. lineata* and 111 specimens of atrypids were assigned to *Pseudoatrypa* sp. A. Representatives of the two species from this study are shown in Fig. 4.4. The first two principal component axes (axes 1 and 2) explained a total of 50.2 % of the variation in dorsal valves, 52.9 % of the variation in ventral valves, 82.0 % of the variation in the posterior region and 98.0 % of the variation in the anterior region (Fig. 4.2). Principal component analysis of dorsal and ventral valves indicates that there is considerable shape variation within each species and that the two species overlap considerably in the morphology of both valves in the x–y plane (Fig. 4.2e, f). Procrustes distances between the mean shape of the two species are 0.023 for the dorsal valves and 0.028 for ventral valves, suggesting that ventral valves show slightly greater morphological differentiation than dorsal valves. Principal component analysis of posterior and anterior regions also indicates that there is considerable variation in morphology between the two species in the y–z plane (Fig. 4.2g, h). Procrustes distances between the mean shape of the two species are 1.69 for the posterior region and 1.53 for anterior region.

MANOVA detects a small significant difference in mean shape of dorsal, ventral, posterior and anterior regions ($p < 0.01$). DFA detects a small but significant difference in mean shape of dorsal and ventral valves ($p < 0.01$) and a large significant difference in mean shape of the posterior and anterior regions

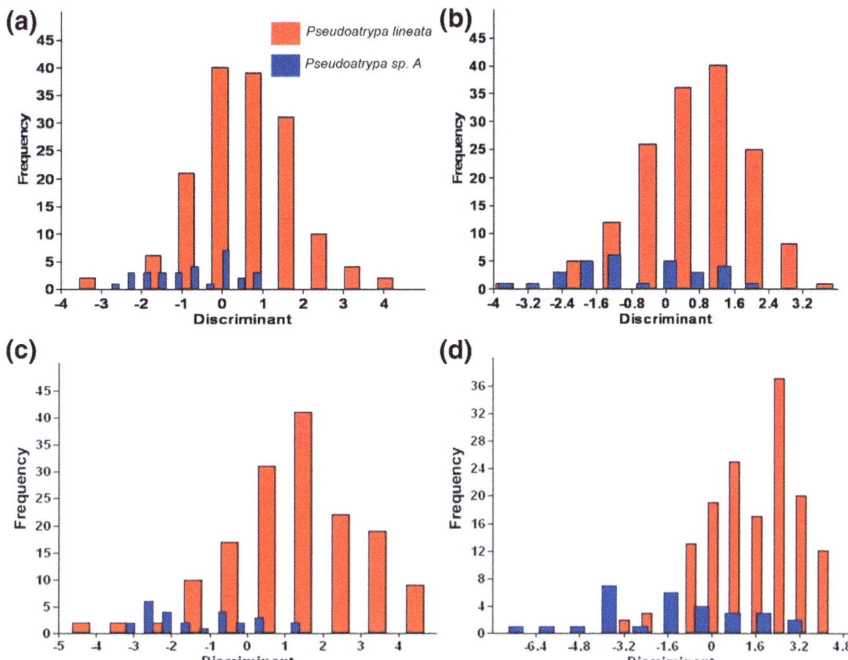

Fig. 4.5 Discriminant function analysis showing the morphological distinctness between *P lineata* and *Pseudoatrypa* sp. A for **a** dorsal valves: Hotelling's t2 P = 0.00759 (Dorsal valves DFA) **b** ventral valves: Hotelling's t2 P = 0.002113 (ventral valves DFA), **c** posterior region: Hotelling's t2 P = 6.304×10^{-9} (posterior region DFA), and **d** anterior region: Hotelling's t2 P = 1.973×10^{-12} (anterior region DFA)

($p < 0.01$) (Fig. 4.5). The significance of the MANOVA and the DFA demonstrates that the two populations can be distinguished as separate species, based on shell shape.

Thin plate spline visualisation plots show the mean morphological shapes of these two species are different (Fig. 4.6). Dorsal valves show a difference in the shape of the posterior hinge line and anterior commissure. The distances between the umbo tip and the posterior left and right lateral margins in the dorsal valve plots are less in *P. lineata* than in *Pseudoatrypa* sp. A, confirming the observation of a more widened hinge line in the latter. Similarly, the distances between the middle point of commissure and the lowest point of the median deflection (fold and sulcus) on both halves of the specimen suggesting a widening of the deflection in *P. lineata* and narrowing in *Pseudoatrypa* sp. A. Ventral valves, however, do not show much significant difference in shape, except for the widened hinge line in *Pseudoatrypa* sp. A relative to the narrow hinge line in *Pseudoatrypa lineata*. The posterior region plots show greater distance between the dorsal umbo and ventral beak tip and lesser distance between dorsal umbo and maximum curvature point in *P. lineata* than in *Pseudoatrypa* sp. A, consistent with the visual observation of a

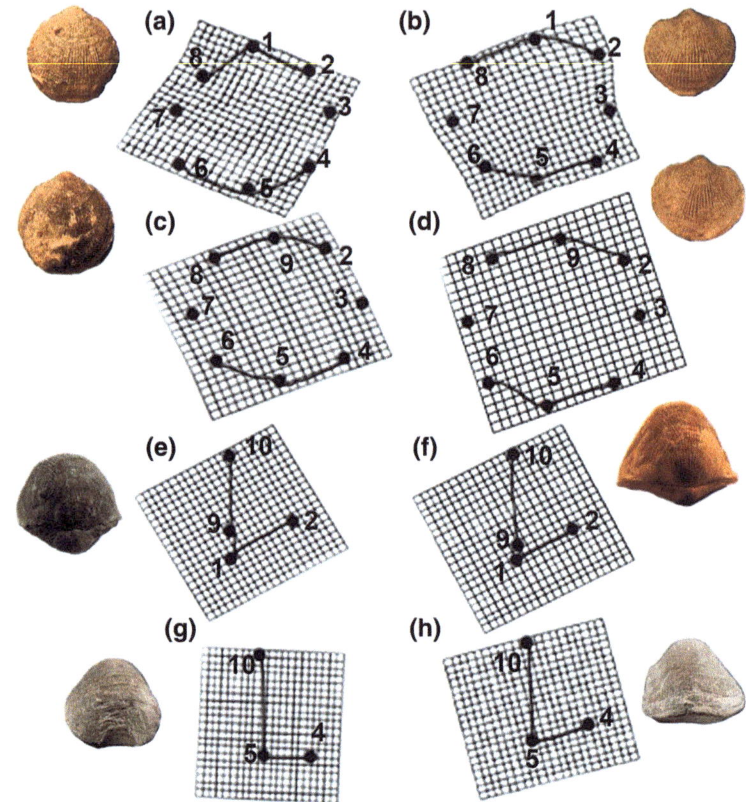

Fig. 4.6 Thin Plate Spline visualisation plots for mean morphological shape of **a** dorsal valves of *P. lineata*, **b** dorsal valves of *Pseudoatrypa* sp. A, **c** ventral valves of *P. lineata*, **d** ventral valves of *Pseudoatrypa* sp. A, **e** posterior region of *P. lineata*, **f** posterior region of *Pseudoatrypa* sp. A, **g** anterior region of *P. lineata* and **h** anterior region of *Pseudoatrypa* sp. A

domal, relatively shallower dorsal valve and inflated ventral valve in *P. lineata* and arched, relatively deeper dorsal valve and flattened ventral valve in *Pseudoatrypa* sp. A. The mean shape plots for the anterior region show less distance between the mid-anterior and right anterior margin in *P. lineata* than *Pseudoatrypa* sp. A and the lateral margin is higher in *Pseudoatrypa* sp. A than in *P. lineata*. This demonstrates that the two species are substantially different in morphological shape (Fig. 4.6).

4.4.2 Frequency of Episkeletobionts

Pseudoatrypa lineata and *Pseudoatrypa* sp. A are hosts to many colonial episkeletobionts—hederellids, sheet-like bryozoans, tabulate corals, fenestrate bryozoans and *Ascodictyon*—as well as many solitary episkeletobionts—

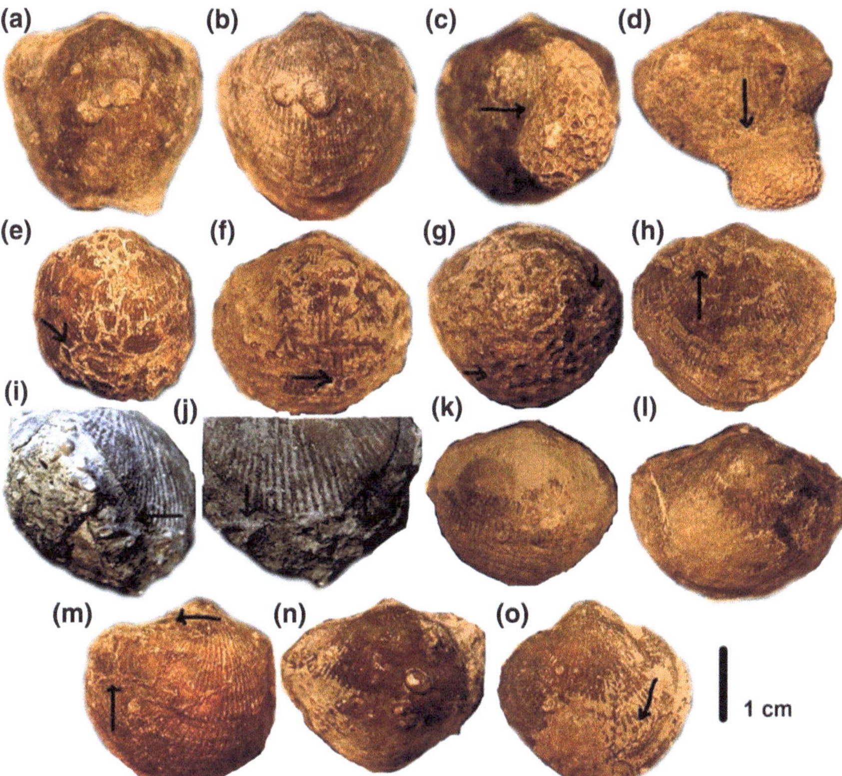

Fig. 4.7 Different types of episkeletobionts on *P. lineata* and *Pseudoatrypa* sp. A hosts—**a**, **b** (IU#100196, IU#100211) Microconchid tube-worms; **c, d** (IU#100179, IU#100122) tabulate sheet corals; **e–j** (IU#100061, IU#100109, IU#100164, IU#100177, IU#100241) auloporid coral colonies; **k** (IU#100222) craniid brachiopod; l) (IU#100174) bryozoans sheet; m) (IU#100077) hederellid colony; **n, o** (IU#100138, IU#100226) mutual co-occurrences of hederellid, bryozoan sheet and microconchid tube worms. *Black arrows* indicate the episkeletobiont extension to the posterior or anterior edges of the host valve

microconchids, craniid brachiopods, and *Cornulites* (Fig. 4.7). Episkeletobionts encrusted 155 specimens (out of 232 total specimens) of *P. lineata* and 30 specimens (out of 111 total specimens) of *Pseudoatrypa* sp. A, for a total of 185 encrusted specimens. Episkeletobionts encrusted more frequently on *P. lineata* than *Pseudoatrypa* sp. A. On *Pseudoatrypa lineata*, 125 dorsal valves (81 %) and 65 ventral valves (42 %) out of 155 encrusted specimens were encrusted, compared with 30 dorsal valves (100 %) and 22 ventral valves (74 %) out of 30 encrusted specimens of *Pseudoatrypa* sp. A.

A total of 354 episkeletobionts encrusted *P. lineata* dorsal valves ($A_c = 2.83$ %) and 74 episkeletobionts encrusted ventral valves ($A_c = 1.14$ %) On *Pseudoatrypa* sp. A, 152 episkeletobionts encrusted dorsal valves ($A_c = 5.07$ %) and 61 encrusted ventral valves ($A_c = 2.77$ %) Dorsal valves are more heavily encrusted for both

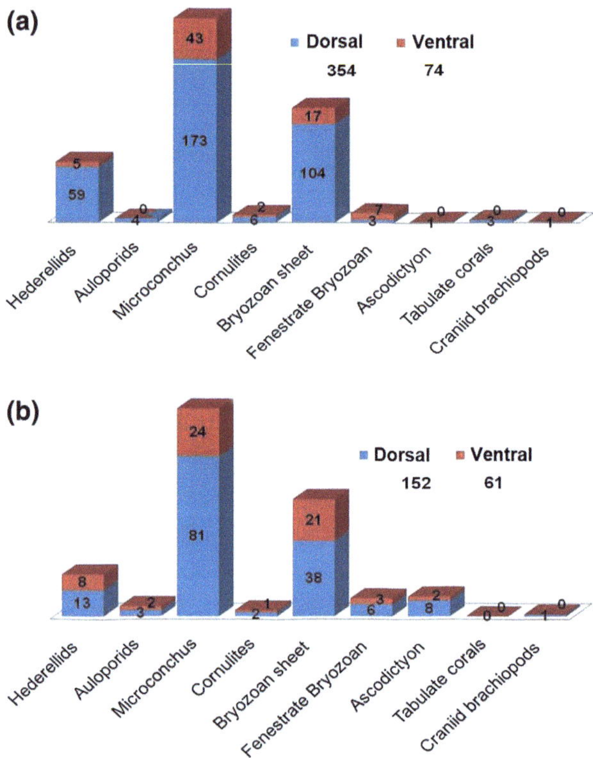

Fig. 4.8 Total episkeletobiont count on **a** *P. lineata* and **b** *Pseudoatrypa* sp. A

Table 4.1 Mean encrustation frequency (A_C) results of each episkeletobiont type on *P. lineata* and *Pseudoatrypa* sp. A

Episkeletobiont type	*Pseudoatrypa lineata* (dorsal)	*Pseudoatrypa lineata* (ventral)	*Pseudoatrypa* sp. A (dorsal)	*Pseudoatrypa* sp. A (ventral)
Hederellids	**0.3**	0.05	**1.44**	1.21
Auloporids	0.02	0	0.33	0.3
Microconchus	**0.89**	0.43	**9.00**	3.64
Cornulites	0.03	0.02	0.22	0.15
Bryozoan sheet	**0.54**	0.17	**4.22**	3.18
Fenestrate Bryozoan	0.02	0.07	0.67	0.45
Ascodictyon	0.01	0	0.89	0.3
Tabulate corals	0.02	0	0	0
Craniid brachiopods	0.01	0	0.11	0

Bold values indicate greater encrustation frequency

Table 4.2 Summary of encrustation by episkeletobiont type. The number of brachiopods encrusted by each episkeletobiont (shells encrusted) and the percentage of encrusted shells that had that particular episkeletobiont (%) are reported

Episkeletobiont	*Pseudoatrypa lineata*					*Pseudoatrypa* sp. A					Grand totals	
	Shells encrusted	%	Dorsal encrusters	Ventral encrusters	P	Shells encrusted	%	Dorsal encrusters	Ventral encrusters	P	shells	Encrusters
Hederellids	23	14.8	59	5	**<0.05**	5	16.7	13	8	**<0.05**	28	85
Auloporids	4	2.6	4	0		2	6.7	3	2		6	9
Microconchus	55	35.5	173	43	**<0.05**	10	30.0	81	24	**<0.05**	64	321
Cornulites	2	1.3	6	2		2	6.7	2	1		4	11
Bryozoan sheet	64	41.3	104	17	**<0.05**	6	20.0	38	21	**<0.05**	70	180
Fenestrate Bryozoan	3	1.9	3	7		2	6.7	6	3		5	19
Ascodictyon	1	0.7	1	0		2	6.7	8	2		3	11
Tabulate corals	2	1.9	3	0		0	0.0	0	0		3	3
Craniid brachiopods	1	0.7	1	0		1	3.3	1	0		2	2
Total	155		354	74		30		152	61		185	641

The total number of encrusters of each episkeletobiont are reported for each valve. *p* values report the probability that the rate of encrustation is the same on dorsal and ventral valves. Grand totals give the total number of shells encrusted by each episkeletobiont in both species and the total number of encrustations by each episkeletobiont. Bold values indicate significant P values suggesting significant difference between frequency of dorsal and ventral encrusters for particular episkeletobiont types

Table 4.3 Sum of the observed and expected number of episkeletobiont activityacross six shell regions of (a) dorsal valves and (b) ventral valves of *Pseudoatrypa lineata* and *Pseudoatrypa* sp. A

DORSAL	*Pseudoatrypa lineata*		*Pseudoatrypa* sp. A	
Regions	Observed	Expected	Observed	Expected
PLL	68.4	65.0	32.584	32.544
PM	65.7	49.8	46.25	24.26
PRL	61.9	65.0	22.08	32.54
ALL	28.7	53.1	4.333	19.43
AM	68.4	75.1	20.17	24.41
ARL	68.0	53.1	25.25	18.38
SUM	361.1	361.1	150.667	151.564
VENTRAL	*Pseudoatrypa lineata*		*Pseudoatrypa* sp. A	
Regions	Observed	Expected	Observed	Expected
PLL	15	9.18	11	6.696
PM	6	7.038	7	5.006
PRL	10	9.18	4	6.696
ALL	2	7.497	2	3.782
AM	9	10.608	4.83	5.022
ARL	9	7.497	2.17	3.782
SUM	51	51	31	30.984

The six regions are as follows: *PLL* postero–left lateral region; *PM* posteromedial region; *PRL* postero–right lateral region; *ALL* antero–left lateral region; *AM* anteromedial region; *ARL* antero–right lateral region. Note that in ventral valves, many rare episkeletobionts were absent, so only the most abundant episkeletobionts is considered

species (Fig. 4.8, Table 4.1). However, average encrustation frequency (A_A) was only weakly correlated with the principal component (PC1) scores for both *P. lineata* (dorsal view: r = −0.08, p = 0.36; posterior view: r = 0.08, p = 0.40; anterior view: r = 0.01, p = 0.87) and *Pseudoatrypa* sp. A (dorsal view: r = 0.09, p = 0.63; posterior view: r = −0.06, p = 0.75; anterior view: r = −0.10, p = 0.62), implying that episkeletobionts did not have a strict preference for shape.

Microconchids, hederellids and sheet bryozoans were the most abundant epizoans, while tabulate corals, auloporid corals, craniid brachiopods, fenestrate bryozoans, *Cornulites* and *Ascodictyon* were present but rarer. Overall, dorsal valves of both species were encrusted more frequently by microconchids, sheet bryozoans and hederellid colonies (*Chi-square* test, *p < 0.05*) (Table 4.2). Frequencies of each episkeletobiont taxon on both valves of the two species are illustrated in Table 4.2 and Fig. 4.8. For each episkeletobiont taxon, mean frequency based on encrustation count (A_C) is presented in Table 4.1.

Table 4.4 Results of *Chi-square* test of observed versus expected episkeletobiont activity for *P. lineata and Pseudoatrypa* sp. A (a) dorsal and (b) ventral valves

A) *Pseudoatrypa lineata*	*p* values	Observed is—than expected
PLL versus sum of other areas	0.6410	
PM versus sum of other areas	0.0150	More
PRL versus sum of other areas	0.6710	
ALL versus sum of other areas	0.0003	Less
AM versus sum of other areas	0.3850	
ARL versus sum of other areas	0.0270	More
A) *Pseudoatrypa* sp. A	*p* values	Observed is—than expected
PLL versus sum of other areas	0.9300	
PM versus sum of other areas	9.4×10^{-7}	More
PRL versus sum of other areas	0.042	Less
ALL versus sum of other areas	0.0003	Less
AM versus sum of other areas	0.3639	
ARL versus sum of other areas	0.0822	More
B) *Pseudoatrypa lineata*	*p* values	Observed—than expected
PLL versus sum of other areas	0.0300	More
PM versus sum of other areas	0.6700	
PRL versus sum of other areas	0.5400	
ALL versus sum of other areas	0.0300	Less
AM versus sum of other areas	0.5800	
ARL versus sum of other areas	0.5500	
B) *Pseudoatrypa* sp. A	*p* values	Observed is—than expected
PLL versus sum of other areas	0.0600	
PM versus sum of other areas	0.3300	
PRL versus sum of other areas	0.2400	
ALL versus sum of other areas	0.3200	
AM versus sum of other areas	0.9200	
ARL versus sum of other areas	0.3700	

A *p* value < 0.05 indicates either more or less biological activity in that shell region than expected, as indicated

4.4.3 *Location of Episkeletobionts*

The frequency of biotic interactions varies among the six regions on both valves of the *P. lineata* and *Pseudoatrypa* sp. A hosts; dorsal valves of each are illustrated in Figs. 4.9 and 4.10. Dorsal valves are more heavily encrusted than the ventral valves with relatively greater episkeletobiont concentration on all the grids ($p < 0.01$).

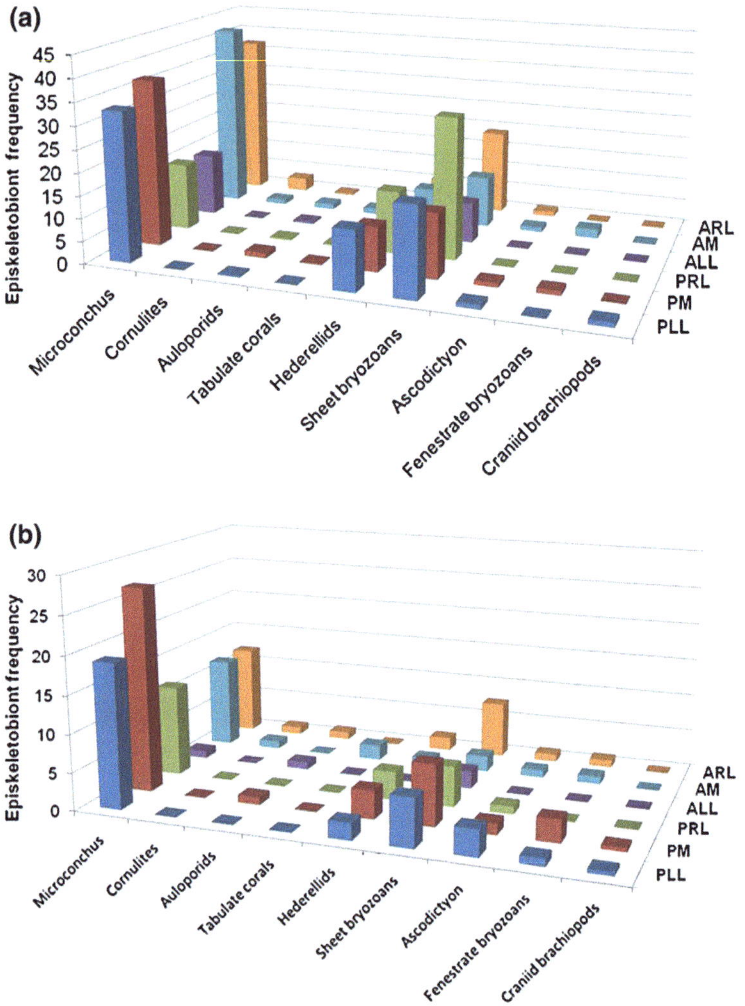

Fig. 4.9 Total standardized frequency of each episkeletobiont activity on dorsal valves across each region for **a** 125 *P. lineata* and **b** 30 *Pseudoatrypa* sp. A hosts. (Note: Standardized frequency = Frequency of colonized episkeletobionts on host species). The six regions are as follows: *PLL* postero–left lateral region; *PM* posteromedial region; *PRL* postero–right lateral region; *ALL* antero–left lateral region; *AM* anteromedial region; and *ARL* antero–right lateral region

4.4.3.1 Dorsal Valves

In dorsal valves, the posteromedial region contains the most frequent occurrence of episkeletobionts on both *P. lineata* and *Pseudoatrypa* sp. A. Microconchids, the most frequent episkeletobiont, is noticeably abundant on all of the six shell regions of both species (Fig. 4.9).

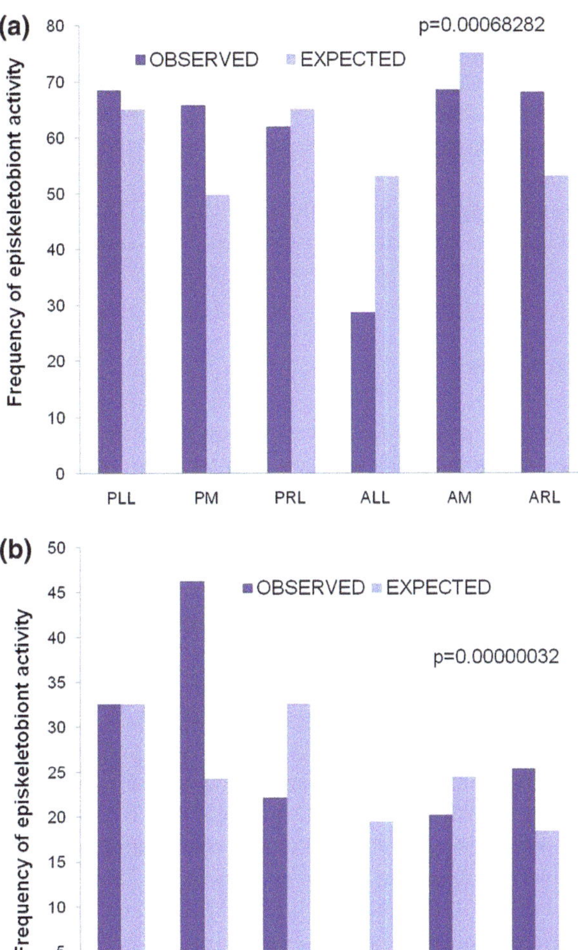

Fig. 4.10 Observed versus expected episkeletobiont activity in dorsal valves across each region. Observed values are the actual frequency of encrustation for each region of the shell; expected values are calculated as described in the text. **a** 125 count *P. lineata* hosts and *b* 30 count *Pseudoatrypa* sp. A hosts. The six regions are as follows: *PLL* postero–left lateral region; *PM* posteromedial region; *PRL* postero–right lateral region; *ALL* antero–left lateral region; *AM* anteromedial region; and *ARL* antero–right lateral region

The episkeletobiont distribution was non-random on dorsal valves. The observed frequency of total episkeletobionts across all regions of the two *Pseudoatrypa* species is significantly different than expected if episkeletobionts randomly encrusted any portion of the shell (*Chi-square*; $p \ll 0.01$). (Table 4.3). Specifically, the antero-left lateral region of *P. lineata* was encrusted at a lower rate than expected (*Chi-square, p* $\ll$ 0.01) whereas the postero-right lateral and antero-left lateral regions were encrusted at a lower than expected rate on *Pseudoatrypa* sp. A (*Chi-square, p* = 0.04, p $\ll$ 0.01) respectively (Table 4.4, Fig. 4.3). Conversely, the diagonally opposite posteromedial and antero-right lateral regions were encrusted at a greater frequency than expected in both the species (*Chi-square, p* = 0.015 and *p* < 0.027) in *P. lineata; p* $\ll$ 0.01 and *p* = 0.08 in *Pseudoatrypa* sp. A) (Table 4.4, Fig. 4.3). The remaining regions do not show any significant difference between expected and observed episkeletobiont frequency (Table 4.4, Fig. 4.10).

4.4.3.2 Ventral Valves

On ventral valves, the postero-left lateral region bears abundant episkeletobionts on both species. Microconchids are common to all of the six shell regions of

P. lineata, but absent in postero-right lateral and antero-right lateral regions of *Pseudoatrypa* sp. A.

The ventral valve episkeletobiont distribution was too low to infer whether the distribution was random or non-random in the six regions. The observed frequency of total episkeletobionts across all regions of the two *Pseudoatrypa* species for ventral valves is not significantly different than expected (*Chi-square, P. lineata, p* = 0.13 and *Pseudoatrypa* sp. A, *p* = 0.29) (Table 4.3). However, the postero-left lateral region of *P. lineata* was encrusted at a greater rate than expected (*Chi-square, p* = 0.03) and the antero-left lateral region was encrusted at a lower rate than expected rate on *P. lineata* (*Chi-square, p* = 0.03) (Table 4.4, Fig. 4.3). The remaining regions do not show any significant difference between expected and observed episkeletobiont frequency on *P. lineata* (Table 4.4). None of the six regions of the ventral valve show any significant difference between expected and observed episkeletobiont activity in *Pseudoatrypa* sp. A (Table 4.4).

4.5 Discussion

4.5.1 *Morphology of* Pseudoatrypa lineata *and* Pseudoatrypa *sp. A*

The two species *P. lineata* and *Pseudoatrypa* sp. are different enough to warrant splitting into separate species. These samples show significant morphological differences, especially in their dorsal convexity, that are sufficient to designate them as two distinct species based on qualitative traits and significant morphometric shape. Thus, in addition to the visually distinguishing characteristic differences in shell shape, morphometric results suggest that the two morphotypes deserve species-level distinction. In this study, morphological shape variation exists within each species. Principal component analysis indicates morphologies of the two species overlap when dorsal and ventral valves are considered. However, the morphological differences between the two species is best observed in the posterior and anterior views, which is also observed in overall qualitative traits of these species (Figs. 4.2, 4.4, 4.5). The morphological differences between the two species are clearly visible in the thin plate spline plots and the differences in mean are significant using MANOVA, when the posterior and anterior regions are assessed. The two species are separable by DFA (Fig. 4.5). These differences were in the shape of the hinge line and commissure, height of dorsal valve curvature, and ventral valve inflation (Figs. 4.4, 4.5). Thus, we consider the two morphotypes to be different species.

Pseudoatrypa lineata (Webster 1921), was described by Fenton and Fenton (1935) and Day and Copper (1998) as a medium-large sized atrypide with dorsibiconvex to convexoplanar shells with inflated dome-like dorsal valves. *P. lineata* from Cedar Valley of Iowa possessed fine radial, subtubular to tubular rib structure (1–2/1 mm at anterior margin), irregularly spaced concentric growth lamellae (more like wrinkles or lines) crowding towards the anterior and lateral margins in larger adults (20 mm length), with very short projecting frills or almost absent. Day and Copper (1998) grouped *Atrypa lineata*, a growth variant form of *A. lineata* and a subspecies of this species, *A. lineata var. inflata* (described earlier by Fenton and Fenton (1935)), all under *P. lineata*. In our study, one of the species sampled from the Givetian age Traverse Group resemble the overall shape and morphology of *P. lineata* described previously from the late Givetian Cedar Valley of Iowa by Fenton and Fenton (1935) and hence are called *P. lineata* for the purpose of this study.

In contrast, the other species do not resemble *P. lineata* type specimens in that they possess a highly arched dorsal valve in contrast to the dome shaped dorsal valve, a flat ventral valve in contrast to the slightly inflated ventral valve, coarser ribs in contrast to the fine-medium ribs, and a subquadrate shell outline in contrast to the rounded outline of *P. lineata*. *P. devoniana* has an elongated shell outline and is Late Frasnian in age (Day and Copper 1998). *P. witzkei* has a rounded shell outline, a shallow dorsal valve profile and is middle Frasnian in age (Day and Copper 1998). *Desquamatia (Independatrypa) scutiformis* has a subrounded shell outline, shallow dorsal valve and highly imbricate tubular ribs in contrast to the subtubular ribs in this species. Thus, the species described in this study is referred to *Pseudoatrypa* sp. A, as it does not resemble *P. lineata* or other species of late Givetian time.

P. lineata diagnosed in this study clearly resembles *A. lineata var. inflata* of Fenton and Fenton (1935) in having similar shell size (2.1–2.4 ± 0.2 cm), shell thickness and convexity (slightly convex ventral valve), ribs with implantations and bifurcations, and numerous growth lines crowding at the anterior. However, *Pseudoatrypa* sp. A resembles *A. lineata* described by Fenton and Fenton (1935) in having larger sized shells (2.3–3.3 ± 0.2 cm), flattened ventral valve, etc. but is significantly different from *A. lineata* in its dorsal valve shape. Thus, both

P. lineata and *Pseudoatrypa* sp. A in this study are described as two distinct species of *Pseudoatrypa* (Fig. 4.4) based on ornamentation and overall shell shape differences (Figs. 4.4, 4.5, 4.6).

Several studies have speculated possible causes (e.g., sedimentation rates, current stimuli, oxygen level, substrate conditions, etc.) behind the observed morphologies for brachiopod shell shape (Lamont 1934; Bowen 1966; Copper 1967; Alexander 1975; Richards 1969, 1972; Leighton 1998). Copper (1973) suggested *Pseudoatrypa* was a soft muddy bottom inhabitant favoring quieter water, though in later studies (Day 1998; Day and Copper 1998), *Pseudoatrypa* has been reported from carbonate environments of the Upper Devonian Cedar Valley group of Iowa. In this study, the samples of the two species were collected from a thin succession of argillaceous shales representing middle shelf environments. These argillaceous beds contained large atrypids with encrusters attached on both valves, which suggests that these shells might have been subject to agitated currents from time to time that were capable of occasionally flipping the shells over in a moderate-low water energy conditions, thus enabling the growth of encrusters on both sides. Thus, as the *P. lineata* and *Pseudoatrypa* sp. A both existed in the same sedimentological regime and were exposed to similar environmental conditions (similar oxygen-level, energy, and substrate conditions), the causes of the difference in shape of the two species are more likely genetic rather than ecomorphic.

4.5.2 Species Preference of Episkeletobionts

Episkeletobionts more frequently encrusted *P. lineata* over *Pseudoatrypa* sp. A (Tables 4.1, 4.2.). In particular, microconchids and sheet bryozoans, the two most abundant episkeletobionts, were more common on *P. lineata* than *Pseudoatrypa* sp. A. No strict species preference was observed for the third-most abundant taxon, hederellids, or for any of the rarer episkeletobionts (*Cornulites*, auloporid corals, tabulate corals, fenestrate bryozoans, *Ascodictyon*, craniid brachiopods).

Among the abundant episkeletobionts, the greatest episkeletobiont activity on the two species can be attributed to the calcareous spirorbiform microconchid tube worms. These spiral worm tubes (0.5–5.0 mm) on *Pseudoatrypa* hosts resemble those encrusting the Middle Devonian Hamilton Group brachiopod hosts (Bordeaux and Brett 1990 Fig. 4.2). A few tube worms, as shown in Fig. 4.7, were very large (>3 mm).

Bryozoan sheets, the second most abundant episkeletobiont, generally encrusted *Pseudoatrypa* hosts over large surface areas (Fig. 4.7l, n, o). They were more common on *P. lineata* than *Pseudoatrypa* sp. A. The surface covering patterns and colonial morphology of the trepostome bryozoans was similar to those encrusting Middle Devonian brachiopods from the Kashong Shale (Bordeaux and Brett 1990). In a few instances, bryozoans even overgrew spirorbiform microconchids and hederellids (Fig. 4.7n, o).

Hederellids, the third most abundant episkeletobiont, were common on both species and did not show a strict preference for either host. Hederellids, originally defined as suborder Hederolloidea (Bassler 1939), are characterized by tubular, calcitic branches. Hederellids have been traditionally referred to bryozoans, but the true affinity of hederelloids has been called into question by some recent workers (Wilson and Taylor 2001; Taylor and Wilson 2008). These Genshaw Formation hederellids (Fig. 4.7m, o) resemble the hederellid species *Hederella canadensis* that encrusted brachiopods from the Devonian Silica Formation of northwestern Ohio (Hoare and Steller 1967; Pl. 1), brachiopods from the Middle Devonian Kashong Shale of New York (Bordeaux and Brett 1990, Fig. 4.2) and *Paraspirifer bownockeri* from the Michigan Basin (Sparks et al. 1980).

Coarse ribs and spines on brachiopod shells have been considered anti-predatory and anti-fouling tools (Richards and Shabica 1969; Vermeij 1977; Alexander 1990; Leighton 1999, 2003; Carrera 2000; Dietl and Kelley 2001; Schneider 2003, 2009a; Schneider and Leighton 2007; Voros 2010) and in some cases have been avoided by epizoans (Richards 1972). Consistent with this hypothesis, finer ribbed taxa in modern (Rodland et al. 2004) and Devonian (Hurst 1974; Thayer 1974; Anderson and Megivern 1982; Schneider and Webb 2004; Zhan and Vinn 2007; Schneider 2009b) brachiopod assemblages experienced greater encrustation frequency than more coarsely ribbed taxa. In this study, there may be a similar preference for finer ribs. As there is no relationship between shell shape and encrustation, but episkeletobionts did prefer *P. lineata*, the data suggest that ornamentation may have been the determining factor in encruster preference for hosts. The fine-medium rib structure of *P. lineata* may have attracted more episkeletobionts than the coarsely ribbed *Pseudoatrypa* sp. A specimens.

Thus, although most episkeletobionts do not exhibit a preference for one of the species, the microconchids, and sheet bryozoans clearly exhibited a preference for *P. lineata*. Surprisingly, *Pseudoatrypa* sp. A, which possesses a relatively larger shell size than *P. lineata*, is less preferred by the most abundant episkeletobionts. One possible explanation for this could be the greater surface area provided by the inflated geometry of the dorsal valve of *P. lineata*, despite its smaller overall shell size. In other words, *P. lineata* may have facilitated heavier encrustation by providing a larger surface area for settlement.

4.5.3 Live-Dead and Live–Live Associations

Both live and dead hosts could be used as a substrate for episkeletobiont settlement. While live hosts might attract encrusters through their feeding currents, dead hosts, obviously, can only be used as a substrate for encrustation.

For some episkeletobionts, hosts would serve merely as hard substrates. In an epizoan ecology study performed by Watkins (1981), it was observed that some epizoans had a very weak preference for live hosts. There is evidence of hederellids and sheet bryozoa encrusting dead hosts or wood (Thayer 1974). There is also evidence of microconchids encrusting dead brachiopod hosts from the Upper Devonian Cerro Gordo Member of the Lime Creek Formation of Iowa (Anderson and Megivern 1982). In contrast, in other studies (e.g., Ager 1961; Hoare and Steller 1967; Richards 1974; Kesling and Chilman 1975; Morris and Felton 1993), auloporids, hederellids and cornulitids frequently displayed preferential growth along or toward the commissure, particularly on Devonian alate spiriferides and large atrypides, possibly in order to take advantage of feeding and respiration currents actively generated by the host's lophophore.

Because post-mortem encrustation cannot be ruled out, it is critical to interpret whether the brachiopod host was alive concurrently with the episkeletobionts. We observed that in rare cases, episkeletobionts oriented themselves on the brachiopod host in specific directions or encrusted particular regions to benefit from feeding currents (Fig. 4.7). These instances may indicate live–live associations.

Sparks et al. (1980), suggested a commensal relation between *Paraspirifer bownockeri* hosts and the spirorbiform microconchids, whereas Barringer (2008) noted no preferred location or orientation of microconchids on host valves, and suggested that they simply infested the hard substrates of the brachiopods. In the present study, microconchids randomly encrusted both valves (Fig. 4.7a, b), with no particular concentration along the commissure area nor any particular orientation of their apertures, thus leaving the host-episkeletobiont relationship ambiguous. Their random orientation on the host valves may indicate that they fed from ambient water currents, rather than requiring currents induced by live brachiopods, a result consistent with other studies (Ager 1961; Pitrat and Rogers 1978; Hurst 1974; Kesling et al. 1980; Fagerstrom 1996). Encrusting bryozoans rarely indicate live–live episkeletobiont-host interactions (Fagerstrom 1996). Microconchids or sheet bryozoa in this study could have encrusted the two species whether live or dead, possibly because of the availability of their hard substrate.

Hederellid colonies that encrusted brachiopods with their apertures oriented towards the anterolateral commissure may have been in that position to benefit from host exhalant currents as described by Bordeaux and Brett (1990). *Hederella* has been reported to have a commensal relationship with its host by Sparks et al. (1980), whereby the episkeletobiont benefited from the hard surface for attachment and from the elevation above the soft substrate but this would be also true for a dead host. In our study, orientation of the apertures of hederellid colonies towards the postero-left lateral end of the host and their termination towards the lateral margin could indicate that the hederellid was taking advantage of host-induced currents (Fig. 4.7m). Such an orientation of hederellids indicates that these organisms may have either benefited from host's feeding inhalant currents, and possibly may have harmed the host by taking in too much of the host's food supply, thus implying a commensal, or parasitic relationship. In one particular instance, however, hederellid colonies were found to parallel the anterior commissure, which would support a commensal relation with the host (Fig. 4.7o). Thus, some of these associations of hederellids and host brachiopods provide evidence of a real, biological interaction.

Some of the rare episkeletobionts also possibly encrusted live hosts. Although auloporid corals were rare, four colonies grew from the medial region of the host towards the anterior commissure suggesting the possibility of mutualism in which corals would benefit from the host feeding currents and protect the host from predators by using possible stinging cells (Fig. 4.7e). Another colony grew parallel to its host's commissure, suggesting a commensal relationship in which the encruster may have benefited from host feeding currents (Fig. 4.7g). On another specimen (Fig. 4.7h), the hinge-proximal location of the corallites with their termination towards the postero-left lateral end of the host suggests possible live–live association with a mutualistic relation between the host and the episkeletobiont. In one instance, the tabulate coral colony, located along the anterior commissure and extending upright suggests a possible live–live association (commensal relation) between the host and the episkeletobiont, where the encruster may have benefited from host exhalant currents (Fig. 4.7d). *Cornulites* attached along the anterior commissure of one host, suggesting encrustation of a live host. Based on the encruster location preference, these instances suggest a live–live association between the host and the episkeletobiont, thus providing further evidence of real, biological interactions. For other host-encruster associations (*Ascodictyon*, fenestrate bryozoans, craniid brachiopods), the live or dead status of the host remains unknown.

Overall, the preferred location of the most abundant episkeletobionts (microconchids, sheet bryozoans and hederellids) and the rare episkeletobionts that show possible evidence of live–live associations (*Cornulites*, auloporid corals, and tabulate corals)is for the posterior region of both dorsal and ventral valves, regardless of host species (described in next section), suggesting that larval settlement of the episkeletobionts may have occurred at the highest point of inflation on brachiopod valves, regardless of whether the brachiopod was dead or alive.

4.5.4 Distribution of Episkeletobionts on P. lineata and Pseudoatrypa sp. A

Distribution of episkeletobionts has assisted in interpretations of life orientation of brachiopod hosts in the past (Hurst 1974). Encrustation frequency (A_C) on hosts assessed from encrustation count (Table 4.2) were preferentially greater for dorsal valves than ventral valves for both species. This suggests that the dorsal valves probably facilitated greater encrustation due to their domal shell geometry relative to the flattened ventral valve, or possibly because the dorsal valve was "up" and so more exposed to settlement by encrusters.

Episkeletobionts may have settled randomly or non-randomly on the host surface of both valves. In this study, episkeletobiont occurrence varies among the six regions sampled on the dorsal valve of the hosts of the two species (Fig. 4.10).

On dorsal valves, microconchids had no preference for posterior or anterior, sheet bryozoans preferred the posterior region over the anterior, and hederellids also preferred the posterior region. Microconchids had a high preference for postero-medial and second-most preference for the antero-medial region on dorsal valves of both species and a high preference for postero-left lateral region and second-most preference for postero-medial region on ventral valves. Sheet bryozoans frequently encrusted the postero-right lateral region of the dorsal valve of both species with a secondary coverage on postero-left lateral, postero-right lateral and antero-right lateral regions (Fig. 4.9). Sheet bryozoans frequented the postero-left and postero-right lateral regions and secondarily encrusted the anteromedial and antero-right lateral regions of the ventral valves. Hederellids often occupied a vast area along the brachiopod host, often along the posterior edge (Figs. 4.7m, o) in both hosts. Although the ventral valve had a comparatively lower concentration of episkeletobionts as compared to dorsal valves, the most abundant episkeletobionts showed similar preferences for posterior versus anterior regions on both valves.

The rare episkeletobionts, *Cornulites*, auloporids and tabulate corals, preferred the anterior regions while *Ascodictyon*, fenestrate bryozoans and craniid brachiopods had no strict preference for anterior or posterior regions of dorsal valves. Tabulate corals and craniid brachiopods were absent on ventral valves, and *Ascodictyon*, fenestrate bryozoans, *Cornulites* and auloporid corals were too low in abundance to determine their preference for a specific region on the valve. Cornulitids, where present, dominated the anterior (antero-medial and antero-right lateral) commissural margin of the host valves with their aperture pointing away from the hinge margin. Auloporid corals most frequently grew their branching colonies along the anterior commissure (antero-left lateral and antero-right lateral) of dorsal valves od both host species with some occurrences in the postero-medial region (Fig. 4.9); the same pattern is also observed for ventral valves. This could simply be a preference of auloporids for settling and growing near their hosts' exhalant currents, a suggested phenomenon for other brachiopod-auloporid associations (e.g., Shou-Hsin 1959; Pitrat and Rogers 1978; Alvarez and Taylor 1987; Taylor and Wilson 2003; Zapalski 2005). Tabulate sheet-like corals, craniid brachiopod, *Ascodictyon*, and fenestrate bryozoans were too low in abundance to determine whether they were random or non-randomly distributed along the grids of the dorsal valve of the two species. Overall, the greater abundance of episkeletobionts on the posterior region of dorsal and ventral valves may have been due to the greater chances of encruster larval settlement on the highest point of the shell geometry, simply because they are higher above the substrate and so more likely to be encountered first by settling larvae. Alternatively, episkeletobiont settlers possibly selected those regions to benefit from host feeding currents. This pattern held true for both hederellid colonies and sheet bryozoans. In addition, hederellids may have selected the posterior edges of the host to benefit from host feeding currents. Although fewer in abundance, the preference of auloporid corals, tabulate corals and cornulitids along the anterior region of the hosts suggest that these episkeletobionts may have selected that region possibly to benefit from the host feeding (exhalant) currents. In general, most episkeletobionts preferred

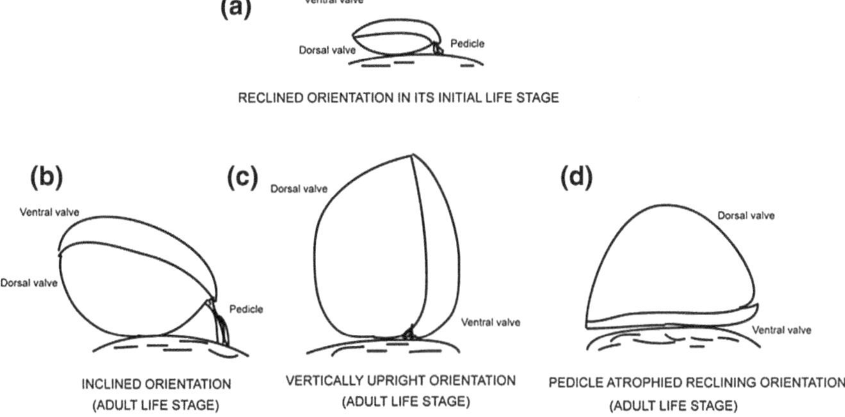

Fig. 4.11 Possible stages of life and death orientation in *P. lineata* and *Pseudoatrypa* sp. A—
a initial immature life stage where the host remain attached by its pedicle to the substrate in a
reclining orientation with the ventral valve up and dorsal valve down, **b** mature adult life stage
where the host remain attached by its pedicle to the substrate in an inclined orientation with the
increasingly convex dorsal valve facing the substrate and the ventral valve facing up, **c** mature
adult life stage where the host remain attached by its pedicle to the substrate in a vertically
upright orientation, **d** mature adult stage with a reclining orientation with dorsal valve up and
ventral valve down after the pedicle has atrophied—this could represent both life and death
orientations in atrypids

margins of posterior and anterior areas of the valves, an encrustation pattern worth
noting. Thus, this nonrandom distribution of episkeletobionts on *Pseudoatrypa* is a
real, biological signal.

4.5.5 *Inference of Life Orientation in* **P. lineata** *and* **Pseudoatrypa** *sp. A*

Atrypids that were dorsibiconvex (dorsal valve more convex than the ventral
valve) to planoconvex (dorsal valve flat and ventral valve convex), lived attached
by a pedicle in their early stages of life in an almost reclining life orientation,
probably with the dorsal valve closest to the substrate and ventral valve facing
upwards (Alexander 1984), and later with increasing convexity of the dorsal valve
through ontogeny, they attained an inclined or a vertically upright life orientation
(Fenton and Fenton 1932) (Fig. 4.11a–c). When the pedicle later atrophied in their
adult stage, they attained a hydrodynamically stable resting life position (Fenton
and Fenton 1932) by falling on their relatively flattened ventral valve with their
convex dorsal valves facing upright into the water column (Fig. 4.11d), such that
the commissure was subparallel to the substrate. In this resting life position, the
roughly domal shape of the shell would provide an optimally streamlined

condition for receiving currents from potentially any direction (Copper 1967). Problematically for interpretation of live–live episkeletobiont-host interactions, this last orientation was also the most likely orientation for the brachiopod after death. In this study, encrustation distribution patterns and frequency suggest that, at the time when most encrustation occurred, these hosts were oriented with their dorsal valves up after pedicle atrophy. Whether these species were alive or dead at the time of encrustation cannot be discerned. Thus, our results from location preference of episkeletobionts on host species alone cannot suggest if these host species were encrusted pre- or post-mortem. In addition, these shells preserved no frills, which would have provided stability to the organism in a particular orientation. Frills, characteristic of atrypid brachiopods, are large growth lamellae that can project beyond the contour of the valves and can assist as anchors in high energy, mobile substrates (Copper 1967). Thus predicting their life orientation without such evidence is difficult based on encrustation distribution pattern alone.

The difference in morphology of the two species poses the question of whether they had similar life orientations. If these species were living in soft substrates, then the shape difference in *P. lineata* (inflated pedicle valve) and *Pseudoatrypa* sp. A (flattened pedicle valve) would have had little or almost no effect in their stability patterns in life or in availability of cryptic surface area, once the brachiopod was in its dorsal-valve-up orientation. However, if these species were resting on hard substrates in the same orientation, then the inflated ventral valves in *P. lineata* would have had a greater surface area exposed along the umbo region for encruster settlement than *Pseudoatrypa* sp. A. Both species would not have had much difference in stability though because of their relatively flattened ventral valve. Given that both these species lived in similar environments, life orientations may also have been similar regardless of whether they lived on soft or hard substrates. Thus, encrustation may have occurred on the shell surface of these species specific to each of their multiple life or death orientations.

4.6 Conclusion

Pseudoatrypa lineata and *Pseudoatrypa* sp. A (Variatrypinae) dominated the atrypide assemblage recovered from the Lower Genshaw Formation of the Middle Devonian Traverse Group of Michigan. The two species were identified based on differences in qualitative traits and statistical shape analysis. *P. lineata* differs from *Pseudoatrypa* sp. A in having a relatively smaller shell size, domal shape with a relatively shallower dorsal valve curvature, slightly convex ventral valve with inflation near the umbo, narrower hinge line, wider commissure with a pronounced gentle to steep fold, and fine-medium sized closely spaced ribs. Statistically significant shape values and large morphological distances between the two species, supports the distinct shapes of the two species identified.

Of the 343 *Pseudoatrypa* hosts examinedfrom both species, 185 of them bore episkeletobionts. The most abundant episkeletobionts were the microconchids, hederellids and the sheet bryozoans. Auloporid corals, *Cornulites*, tabulate corals, *Ascodictyon*, craniid brachiopods, and fenestrate bryozoans were very rare. Several episkeletobionts in this study provide evidence of encrusting a live host based on the location preference of the episkeletobionts. Hederellids, auloporid corals, tabulate corals, and *Cornulites* had a live–live episkeletobiont-host relationship. The majority of other episkeletobionts, notably microconchids, sheet bryozoans, and *Ascodictyon*, were enigmatic in determining whether their relationship was with a live or a dead host. Very few epizoans crossed the commissure of the host after the host's death.

Most episkeletobionts (microconchids and sheet bryozoans) preferred *P. lineata*, despite the fact that this species is generally smaller. This differential effect in epibiosis could be due to the nature of ribbing structure (fine to medium) and greater exposed area facilitated by the shell shape of *P. lineata*. Overall, the episkeletobiont preference for one species over another strongly suggests that the overall episke-letobiont distribution was influenced by shape and ornamentation variation in a-trypid samples. Abundant encrusting organisms—microconchids, sheet bryozoans and hederellids, had a preference for *P. lineata* dorsal valves. This greater abundance of episkeletobionts on dorsal valves and lower abundance on ventral valves is suggestive of most of the encrustation occurring when the host species were oriented with their convex dorsal valves up and ventral valves down with most of the ventral valve surface in contact with the sediment substrate. Whether encrustation was pre- or post-mortem was challenging to discern for the majority of host-episkeletobiont associations as life orientation of the host would also be a hydrodynamically stable orientation of the articulated shell after death. Additionally, the most abundant episkeletobionts showed a preference for the posterior region on both dorsal and ventral valves of both species. This suggests that the posterior umbonal region may have provided an inflated surface that remained exposed, thus, favoring the settle-ment of most episkeletobiont larvae in that region.

The present study of the Genshaw Formation documents epibiosis on two species of atrypids, which significantly enhances our understanding of morpho-logical influence on episkeletobiont distribution.

References

Ager DV (1961) The epifauna of a Devonian spiriferid. Q J Geol Soc Lond 117:1–10

Alexander RR (1975) Phenotypic liability of the brachiopod *Rafinesquina alternata* (Ordovician) and its correlation with the sedimentologic regime. J Paleontol 49:607–618

Alexander RR (1984) Comparative hydrodynamic stability of brachiopod shells on current-scoured arenaceous substrates. Lethaia 17:17–32

Alexander RR (1990) Mechanical strength of shells of selected extant articulate brachiopods: implications for paleozoic morphologic trends. His Biol Int J Paleobiol 3:169–188

Alvarez F, Taylor PD (1987) Epizoan ecology and interactions in the Devonian of Spain. Palaeogeogr Palaeoclimatol Palaeoecol 61:17–31

Anderson WI, Megivern KD (1982) Epizoans from the Cerro Gordo member of the lime creek formation (upper Devonian), Rockford, Iowa. In: Proceedings of the Iowa academy of science, vol 89, pp 71–80

Barringer JE (2008) Analysis of the occurrence of microconchids on middle Devonian brachiopods from the Michigan basin: implications for microconchid and brachiopod autecology. Unpublished M S thesis, Michigan State University, Michigan, pp 1–127

Bartholomew JB (2006) Middle Devonian faunas of the Michigan and Appalachian Basins: comparing patterns of biotic stability and turnover between two paleobiogeographic subprovinces. Unpublished M S thesis, University of Cincinnati, Ohio, pp 1–300

Bassler RS (1939) The Hederelloidea, a suborder of Paleozoic cyclostomatous Bryozoa. In: Proceedings of the United States National Museum, vol 87, pp 25–91

Bookstein FL (1989) Principal warps: thin-plate splines and the decomposition of deformations. IEEE Trans Pattern Anal Mach Intell 11:567–585

Bookstein FL (1991) Morphometric tools for landmark data: geometry and biology. Cambridge University Press, New York

Bordeaux YL, Brett CE (1990) Substrate specific associations of epibionts on middle Devonian brachiopods: implications for Paleoecology. Hist Biol 4:203–220

Bose R, Schneider C, Polly PD, Yacobucci MM (2010) Ecological interactions between *Rhipidomella* (Orthides, Brachiopoda) and its endoskeletobionts and predators from the middle Devonian Dundee Formation of Ohio, United States. Palaios 25:196–210

Bowen ZP (1966) Intraspecific variation in the Brachial Cardinalia of *Atrypa reticularis*. J Paleontol 40:1017–1022

Brett CE, Baird GC, Bartholomew AJ, DeSantis MK, Straeten CA (2010) Sequence stratigraphy and a revised sea-level curve for the middle Devonian of eastern North America. Palaeogeogr Palaeoclimatol Palaeoecol 304:21–53

Brezinski DK (1984) Upper Mississippian epizoans and hosts from southwestern Pennsylvania. In: Proceedings of the Pennsylvania academy of science, vol 58, pp 223–226

Carrera MG (2000) Epizoan-sponge interactions in the early Ordovician of the Argentine Precordillera. Palaios 15:261–272

Copper P (1967) Adaptations and life habits of Devonian atrypid brachiopods. Palaeogeogr Palaeoclimatol Palaeoecol 3:363–379

Copper P (1973) New Siluro-Devonian atrypoid brachiopods. J Paleontol 47:484–500

Curry GB (1983) Brachiopod caeca—a respiratory role? Lethaia 16:311–312

Day J (1998) Distribution of latest Givetian-Frasnian Atrypida (Brachiopoda) in central and western North America. Acta Palaeontologica Polonica 43:205–240

Day J, Copper P (1998) Revision of latest Givetian-Frasnian Atrypida (Brachiopoda) from central North America. Acta Palaeontologica Polonica 43:155–204

Dietl GP, Kelley PH (2001) Mid-Paleozoic latitudinal predation gradient: distribution of brachiopod ornamentation reflects shifting carboniferous climate. Geology 29:111–114

Ehlers GM, Kesling RV (1970) Devonian strata of Alpena and Presque Isle counties, Michigan. Guidebook for field trips, Michigan Basin Geological Society, pp 1–130

Fagerstrom JA (1996) Paleozoic brachiopod symbioses: testing the limits of modern analogues in paleoecology. Geol Soc Am Bull 108:1393–1403

Fenton CL, Fenton MA (1932) Orientation and injury in the genus *Atrypa*. Am Midl Nat 13:63–74

Fenton CL, Fenton MA (1935) Atrypae described by Clement L. Webster and related forms (Devonian, Iowa). J Paleontol 9:369–384

Gibson MA (1992) Some epibiont-host and epibiont–epibiont relationships from the Birdsong Shale Member of the Lower Devonian Ross Formation (West-Central Tennessee, U.S.A.). Hist Biol 6:113–132

Hammer Ø, Harper D (2005) Paleontological data analysis. Blackwell Publishing, Oxford

Haney RA, Mitchell CE, Kim K (2001) Geometric morphometric analysis of patterns of shape change in the Ordovician Brachiopod Sowerbyella. Palaios 16:115–125

Hoare RD, Steller DL (1967) A Devonian brachiopod with epifauna. Ohio J Sci 67:291

Hurst JM (1974) Selective epizoan encrustation of some Silurian brachiopods from Gotland. Palaeontology 17:423–429

Kelly AW, Smith GW (1947) Stratigraphy and structure of traverse group in Afton-Onaway area, Michigan. Bull Am Assoc Petrol Geol 31:447–469

Kesling RV, Chilman RB (1975) Strata and megafossils of the middle Devonian silica formation. Pap Paleontol 8:1–408

Kesling RV, Hoare RD, Sparks DK (1980) Epizoans of the Middle Devonian brachiopod *Paraspirifer bownockeri*: their relationships to one another and to their host. J Paleontol 54:1141–1154

Koch WF (1978) Brachiopod paleoecology, paleobiogeography, and biostratigraphy in the upper middle Devonian of Eastern North America: an ecofacies model for the Appalachian, Michigan, and Illinois basins. Unpublished Ph D thesis, Oregon State University, Oregon, pp 1–311

Lamont A (1934) Brachiopod morphology in relation to environment. Cem Lime Gravel 8: 216–219

Leighton LR (1998) Constraining functional hypotheses: controls on the morphology of the concavo-convex brachiopod *Rafinesquina*. Lethaia 3:293–307

Leighton LR (1999) Possible latitudinal predation gradient in middle Paleozoic oceans. Geology 27:47–50

Leighton LR (2003) Predation on brachiopods. In: Kelley PH, Kowalewski M, Hansen TA (eds) Predator-prey interactions in the fossil record: topics in geobiology, vol 20. Kluwer/Plenum, New York, pp 215–237

Lescinsky HL (1995) The life orientation of concavo-convex brachiopods: overturning the paradigm. Paleobiology 21: 520–551

Macleod N, Forey PL (2002) Morphology, shape, and phylogeny. Taylor and Francis, New York

Morris RW, Felton SH (1993) Symbiotic association of Crinoids, Platyceratid Gastropods, and *Cornulites* in the Upper Ordovician (Cincinnatian) of the Cincinnati, Ohio region. Palaios 8:465–476

Morris RW, Felton SH (2003) Paleoecologic associations and secondary tiering of *Cornulites* on crinoids and bivalves in the Upper Ordovician (Cincinnatian) of southwestern Ohio, southeastern Indiana, and northern Kentucky. Palaios 18:546–558

Pitrat CW, Rogers FS (1978) *Spinocyrtia* and its epizoans in the traverse group (Devonian) of Michigan. J Paleontol 52:1315–1324

Richards RP (1969) Biology and ecology of *Rafinesquina alternata (Emmons)*. Geol Soc Am Abstr Programs 6:41–42 (North Central Section)

Richards RP, Shabica CW (1969) Cylindrical living burrows in Ordovician dalmanellid brachiopod beds. J Paleontol 43:838–841

Richards RP (1972) Autecology of Richmondian brachiopods (Late Ordovician of Indiana and Ohio). J Paleontol 46:386–405

Richards RP (1974) Ecology of the Cornulitidae. J Paleontol 48:514–523

Rodland DL, Kowalewski M, Carroll M, Simoes MG (2004) Colonization of a 'Lost World': encrustation patterns in modern subtropical brachiopod assemblages. Palaios 19:381–395

Rodrigues SC, Simoes MG, Kowalewski M, Petti MAV, Nonato EF, Martinez S, Rio CDJ (2008) Biotic interaction between spionid polychaetes and bouchardiid brachiopods: paleoecological, taphonomic and evolutionary implications. Acta Palaeontologica Polonica 53:657–668

Rohlf FJ (1990) Fitting curves to outlines. In: Proceedings of the Michigan Morphometrics Workshop, pp 167–177

Rohlf FJ, Slice DE (1990) Extensions of the Procrustes method for the optimal superimposition of landmarks. Syst Zool 39:40–59

Rohlf FJ (1999) Shape statistics: procrustes superimpositions and tangent spaces. J Classif 16:197–223

Rohlf FJ, Marcus LF (1993) A revolution in morphometrics. Trends Ecol Evol 8:129–132

Rudwick MJS (1962) Filter-feeding mechanisms in some brachiopods from New Zealand. Zool J Linn Soc (Lond) 44:592–615

Sandy MR (1996) Oldest record of peduncular attachment of brachiopods to crinoid stems, Upper Ordovician, Ohio, USA (Brachiopods; Atrypida: Echinodermata; Crinoidea). J Paleontol 70:532–534

Schneider CL (2003) Hitchhiking on Pennsylvanian echinoids: epibionts on archaeocidaris. Palaios 18:435–444

Schneider CL (2009a) Epibionts on Late Carboniferous through Early Permian echinoid spines from Texas, USA, Echinoderms 2006. In: Proceedings of the 12th international echinoderm conference, Durham, 7–11 August 2006, New Hampshire, U.S.A. CRC Press, Boca Raton, pp 1–679

Schneider CL (2009b) Substrate preferences of Middle and Late Devonian *Hederella* from the midcontinent USA. In: Hageman SJ, Key MM Jr, Winston JE (eds) Bryozoan Studies 2007, Proceedings of the 14th IBA Conference, Boone, North Carolina, July 1–8, 2007, vol 15. Virginia Museum of Natural History Special Publication, Martinsville, pp 295–300

Schneider CL, Leighton LR (2007) The influence of spiriferide microornament on Devonian epizoans. Geol Soc Am Abstr Programs 39:531

Schneider CL, Webb A (2004) Where have all the encrusters gone? Encrusting organisms on Devonian versus Mississippian brachiopods. Geol Soc Am Abstr Programs 36:111

Shou-Hsin C (1959) Note on the palaeoecological relation between *Aulopora* and *Mucrospirifer*. Acta Palaeontologica Sinica 7:502–504

Slice DE (2001) Landmark coordinates aligned by procrustes analysis do not lie in Kendall's shape space. Syst Biol 50:141–149

Sparks DK, Hoare RD, Kesling RV (1980) Epizoans on the brachiopod *Paraspirifer bowneckeri* (Stewart) from the Middle Devonian of Ohio, University of Michigan Museum of Paleontology, Papers on Paleontology, vol 23, pp 1–50

Spjeldnaes N (1984) Epifauna as a tool in autecological analysis of Silurian Brachiopods. The Spec Pap Palaeontol 32:225–235

Stumm EC (1951) Check list of fossil invertebrates described from the middle Devonian traverse group of Michigan: Contributions from the Museum of Paleontology, vol 9. University of Michigan, Ann Arbor, pp 1–44

Sumrall CD (2000) The biological implications of an edrioasteroid attached to a pleurocystitid rhombiferan. J Paleontol 74:67–71

Taylor PD, Wilson MA (2002) A new terminology for marine organisms inhabiting hard substrates. Palaios 17:522–525

Taylor PD, Wilson MA (2003) Palaeoecology and evolution of marine hard substrate communities. Earth Sci Rev 62:1–103

Taylor PD, Wilson MA (2008) Morphology and affinities of hederelloid "bryozoans". In: Hageman SJ, Key MM Jr, Winston JE (eds) Bryozoan Studies 2007: Proceedings of the 14th international bryozoology conference, Boone, North Carolina, July 1–8, 2007, vol 15. Virginia Museum of Natural History Special Publication, Martinsville, pp 301–309

Thayer CW (1974) Substrate specificity of Devonian epizoan. J Paleontol 48:881–894

Vermeij GJ (1977) The Mesozoic marine revolution: evidence from molluscs, predation, and grazing. Paleobiology 3:245–258

Voros A (2010) Escalation reflected in ornamentation and diversity history of brachiopod clades during the Mesozoic marine revolution. Palaeogeogr Palaeoclimatol Palaeoecol 291:474–480

Warthin AS, Cooper GA (1935) New formation names in the Michigan Devonian. J Wash Acad Sci 25:524–26

Warthin AS, Cooper GA (1943) Traverse rocks of Thunder bay region, Michigan. Bull Am Assoc Petrol Geol 27:571–595

Watkins R (1981) Epizoan ecology of the type Ludlow Series (Upper Silurian), England. J paleontol 55:29–32

Webster CL (1921) Notes on the genus *Atrypa*, with description of new species. Am Midl Nat 7:13–26

Wiedman LA (1985) Community paleoecological study of the silica shale equivalent of Northeastern Indiana. J Paleontol 59:160–182

Wilson MA, Taylor PD (2001) "Pseudobryozoans" and the problem of encruster diversity in the Paleozoic. PaleoBios 21:134–135

Wylie AS, Huntoon JE (2003) Log curve amplitude slicing—visualization of log data from the Devonian Traverse Group in the Michigan Basin, U.S. Bull Am Assoc Petrol Geol 87: 581–608

Zapalski MK (2005) Paleoecology of Auloporida: an example from the Devonian of the Holy Cross Mts., Poland. Geobios 38:677–683

Zelditch ML, Swiderski DL, Sheets HD, Fink WL (2004) Geometric morphometrics for biologists, a primer. Elsevier Academic Press, New York

Zhan R, Vinn O (2007) Cornulitid epibionts on brachiopod shells from the Late Ordovician (middle Ashgill) of East China. Estonian J Earth Sci 56:101–108

Chapter 5
Success of Geometric Morphometrics in Deducing Morphological Shape Change Patterns in Paleozoic Atrypids

5.1 Conclusion

5.1.1 Brief Inferences

This section concludes all the major findings accomplished during the course of this study. This study described the effects of evolution, ecology and environment on shell morphology of extinct Paleozoic brachiopods both in large scale and small scale temporal and spatial units. Geometric morphometric methods were used to quantify shell shape which further helped in assessment of (a) morphological shape differences in atrypid brachiopods at the subfamily, genus and species level, (b) correctness of a previously constructed phylogeny of atrypids, (c) evolutionary rates and modes in atrypid generic and species lineages, (d) correlation with environmental factors, (e) influence of episkeletobionts on morphological shape, and (f) derive ecology and life habit of these extinct brachiopods.

The taxonomic arrangement and phylogenetic patterns examined from morphological shape distances complies with Copper's (Copper 1973) phylogenetic tree. Morphologies appear to be constrained in the generic lineages in the ~ 70 m.y. time scale with slow rate of evolution, however, diversifying selection has been acting on them. Morphology in an atrypid species lineage from a ~ 5 m.y. strata, exhibits stasis like patterns in the lower stratigraphic intervals with abrupt change occurring in the uppermost interval. Evolutionary rates are slow to moderate with stabilizing selection acting on the species lineage. The abrupt change in the uppermost occurrence could be due to change in environmental conditions during that time. This observed morphological pattern in the Michigan Basin when compared to the Hamilton Group of the Appalachian Basin, suggests that this pattern is unique to the Traverse Group. Morphological shape and ornamentation has influenced episkeletobiont settlement on Genshaw Formation atrypid brachiopod species. The greater concentration of episkeletobionts on convex dorsal valves suggests

R. Bose, *Biodiversity and Evolutionary Ecology of Extinct Organisms*,
Springer Theses, DOI: 10.1007/978-3-642-31721-7_5,
© Springer-Verlag Berlin Heidelberg 2013

their dorsal valve facing up life orientation. The posterior region was inflated on both valves of atrypid species, thus, facilitating greater exposed area for encrustation.

5.1.2 Atrypid Taxonomy, Evolution and Ecology

The study in Chap. 2 provides evidence for greater morphological shape distances between subfamilies than within a subfamily. This result agrees with the phylogenetic arrangement of Copper (1973); however, within genus morphological distances in time and space is as large as between genera, suggesting within group variation is greater than between group variation in atrypids, so much so, that referring individuals to a genus based on morphological shape alone is challenging. Despite the very small morphological divergences among genera within the three atrypid subfamilies, evolutionary rate and mode indicate that diversifying selection has probably been acting on them. Short term changes in individual genus lineages, gets averaged out when compared with other genera, a pattern similar to loose stasis. Distinct taxonomic entities within Atrypidae, show considerable overlap in morphological shape between them, which is also similar to a case of loose stasis. Thus, morphologies in atrypid brachiopods appear to be conserved to a great extent within the P3 EEU.

Chapter 3 study suggests the evolution of the *Pseudoatrypa cf. lineata* at a slow to moderate rate with morphologies lightly constrained within this species lineage. Samples from the lower stratigraphic units in the Traverse Group reflect stasis-like patterns while the uppermost stratigraphic occurrences shows greater morphological change. This abrupt change in morphology may be attributed to the provinciality in the Michigan Basin section during that time interval. This pattern observed in Traverse Group is however, local in scope, unlike the Hamilton Group (correlatable section of Traverse Group) stratigraphic section, that shows overlap in lowermost and uppermost morphological occurrences. Overall, the qualitative similarity noted in morphologies between samples and considerable overlap in morphological variation observed between samples from successive strata suggests that these samples belong to the same species *P. cf. lineata*. The abrupt deviation of the uppermost occurrences from the mean is suggestive of greater environmental change during that time.

The study presented in Chap. 4 shows that while in a very few cases, live–live associations between host brachiopods and episkeletobionts could be discerned, the abundance of most common episkeletobionts on the convex dorsal valves suggest that these brachiopods were most likely oriented with their dorsal valve facing up and ventral valve facing the substrate during life. Greater encrustation abundance on one species over another suggests that episkeletobiont settlement was influenced by variation in morphological shape and ornamentation. Their preferential settlement on the posterior region suggests that the exposed inflated area of the shells may have facilitated encrustation in those regions.

5.1.3 This Study Supports or Contradicts EEU, EESU and PE Model?

Paleontological data reported in this study supports the EEU model and is in partial agreement with the EESU and the punctuated equilibrium model. Morphological dataset of atrypid subfamilies and genera, show overall morphological stability in lineages within the P3 EEU. Morphological dataset of atrypid species lineage in the Traverse Group EESU shows stasis like patterns in the lowermost and middle stratigraphic units, however, marked by abrupt change in the uppermost unit. Morphology of this atrypid species was relatively static during most of its evolutionary history with change occurring at a later time where an associated event of environmental change occurred during that time, which could perhaps be a speciation event. Thus, the datasets from this study largely complies with the EEU and EESU model and thus, provides a great database for future researchers to compare these results with other Phanerozoic EEU and EESUs. Overall, in a broader perspective, brachiopods served as great tools to test models of EEU, EESU and punctuated equilibrium.

This morphometric dataset may also be of future assistance to researchers interested in phylogenetic reconstruction of the complex atrypid brachiopod group, where morphometric shape may be incorporated as an important character in phylogenetic coding coupled with other external and internal morphological characters.

5.1.4 Evolutionary Ecology in Silurian-Devonian Brachiopods

Ecological interactions are thought to have maintained a static adaptive landscape and prevented both the long-term establishment of exotic invading species and evolutionary change of native species within EEUs and EESUs while when disturbance exceeds the capacity of the ecosystem, ecological crashes occur and evolution proceeds at high rates of directional selection (Morris 1995; Morris et al. 1995). In this study, species exhibited morphological stability in the Silurian and Devonian time periods. Similar stable pattern was observed in a species lineage from the Middle Devonian time which exhibited sudden change in morphology later in its life history. This change in morphology coincided with environmental heterogeneity in the Michigan Basin during that time interval (Dorr and Eschman 1970; Ehlers and Kesling 1970; Wylie and Huntoon 2003). This atrypid species lineage, thus, correlated well with environmental conditions persisting during that time, with change being evident of either an immigration or speciation event. Morphology of these shells also influenced episkeletobiont settlement.

All of these combined observations suggests that morphological shape of extinct Paleozoic atrypid brachiopods correlated well with evolutionary, ecological, and environmental factors, providing further insight into the evolutionary ecology of these extinct organisms.

References

Copper P (1973) New Siluro-Devonian atrypoid brachiopods. J Paleontol 47:484–500

Dorr JA, Eschman DF (1970) Geology of Michigan. The University of Michigan Press, 479 p

Ehlers GM, Kesling RV (1970) Devonian strata of Alpena and Presque Isle counties. Michigan basin Geological Society, Michigan 130 p

Morris PJ (1995) Coordinated stasis and ecological locking. Palaios 10:101–102

Morris PJ, Ivany LC, Schopf KM, Brett CE (1995) The challenge of paleoecological stasis: reassessing sources of evolutionary stability. Proc Natl Acad Sci 92:11269–11273

Wylie AS Jr, Huntoon JE (2003) Log-curve amplitude slicing: visualization of log data and depositional trends in the Middle Devonian Traverse Group. Michigan Basin, United States

MIX
Papier aus verantwortungsvollen Quellen
Paper from responsible sources
FSC® C105338

If you have any concerns about our products,
you can contact us on
ProductSafety@springernature.com

In case Publisher is established outside the EU,
the EU authorized representative is:
**Springer Nature Customer Service Center GmbH
Europaplatz 3, 69115 Heidelberg, Germany**

Printed by Libri Plureos GmbH
in Hamburg, Germany